WONDERS
of the
AIR

Tamra Andrews

Wonders of Nature: Natural Phenomena in Science and Myth

LIBRARIES
U N L I M I T E D
A Member of the Greenwood Publishing Group

Westport, Connecticut • London

Library of Congress Cataloging-in-Publication Data

Andrews, Tamra, 1959–
    Wonders of the air / Tamra Andrews.
      p. cm.—(Wonders of nature)
    Includes bibliographical references and index.
    ISBN 1–59158–105–2 (pbk. : alk. paper)
    1. Meteorology—Popular works. I. Title. II. Series: Wonders of nature
(Libraries Unlimited)
    QC863.4.A445   2004
    551.5—dc22   2004048637

British Library Cataloguing in Publication Data is available.

Library of Congress Catalog Card Number: 2004048637
ISBN: 1–59158–105–2

First published in 2004

Libraries Unlimited, 88 Post Road West, Westport, CT 06881
A Member of the Greenwood Publishing Group, Inc.
www.lu.com

Printed in the United States of America

The paper used in this book complies with the
Permanent Paper Standard issued by the National
Information Standards Organization (Z39.48–1984).

10 9 8 7 6 5 4 3 2 1

Contents

CONTENTS

Illustrations

Preface

Teachers' guides are abundant in the area of meteorology, and lesson plans and activities are easily accessible online. This is a different kind of teachers' guide however. It combines science with myth, and it stresses how human curiosity shaped the history of science as ancient people strove to understand and control the weather. Ancient myths contain vivid imagery of atmospheric phenomena, with powerful storm beasts who hide in water pools and behind clouds and who wreak havoc on the world when they move through the air. Some myths in this guide center on atmospheric phenomena that were viewed as both good and evil powers. Atmospheric forces such as wind and rain have both a creative and a destructive function, and they often engage in battles with opposing forces to keep the world moving through the cycle of seasons. In myths, gods of rain fought demons of drought. By reading these myths we can see that ancient mythmakers understood the interdependence of opposites. A world with rain and no drought destroyed the Earth with flood, and a world with drought and no rain reduced the Earth to ashes.

Wonders of the Air is the second in a series of teachers' guides that attempts to teach science through an intermingling of myth and fact. In examining the nature myths of different cultures, we can see that myth and science are intricately entwined. Also, we can see that people have always had a desire to interpret nature's messages. The myths and legends that arose from this desire served as an early form of philosophy and as documentation of scientific observations. Each chapter in this teachers' guide contains an adaptation of a myth and an explanation of the science behind the myth. Topics for discussion and projects should be useful for lessons in art and creative writing, and they center as much on the culture of the people who told the myth as on the science behind it.

The proliferation of books and Web sites that contain information on meteorology and culture made selection of reference sources difficult. I limited the list of teacher resources in the appendix to books that were particularly helpful in preparing this guide and to general Web sites that contain grade appropriate explanations, lessons, and projects or collections of weather myths. In each chapter I included books and Web sites that should help students complete the projects and a list of additional sources to help teachers guide class discussions. Each citation included in the reading lists is annotated.

Much of this material was gathered while doing research for my encyclopedia of nature myths titled *Legends of the Earth, Sea, and Sky: An Encyclopedia of Nature Myths* (ABC-CLIO, 1998). This book can be found in many schools and public libraries, and the paperback edition, published under the name *Dictionary of Nature Myths: Legends of the Earth, Sea, and Sky* (Oxford University Press, 2000), is available in bookstores. Neither this teachers' guide nor my encyclopedia would be possible were it not for Dr. E. C. Krupp of the Griffith Observatory in Los Angeles, Dr. R. Robert Robbins of the University of Texas at Austin, and the many other scholars whose works I relied on for information. The *StarDate* Web site, the Web site for *Earth and Sky*, and many other Web sites listed in the appendix proved to be invaluable for my research.

I'd like to particularly thank my daughter Cristen Andrews for her help in preparing this guide. Cristen is a photojournalism major at the University of Texas at Austin. She created the graphics that accompany this text, as she did for my previous teachers' guide in this series, *Wonders of the Sky*. Cristen's dedication and support mean the world to me. I'd also like to thank Frank Fox, a librarian at Texas State University, for helping me with research, and my editors at Greenwood Publishing for their support and encouragement.

It is my hope that teachers find this guide useful not only for science lessons but also for lessons in social studies, literature, and art. It is also my hope that students who read these myths and learn the science within them gain an understanding of nature's clockwork, an appreciation of nature's power, and a recognition that people of all cultures and at all times in history shared the human desire to interpret the world around them. From this desire, science was born.

Introduction

Long ago, before people undertook a scientific study of the universe, the forces of nature seemed overwhelming. They still do. Wind, rain, and storms have an amazing ability to renew the Earth or to destroy it. Imagine what it would be like to hear the rumbles of thunder and witness the driving rain and believe that some supernatural being had control over these forces. The myths help us imagine what this was like. People told these myths long ago to explain why the rain fell and why the winds blew. They heard the roar of thunder, and they could only attribute it to the roar of a monstrous beast. They saw ice and snow cover the Earth, and they imagined hideous giants intent on freezing everything in existence. Before people understood the science of these forces, they could only hope that they could appease these spirits before they destroyed the world. People had various ways of controlling the spirits. Ancient people had rituals to convince the spirits to release the rains and rituals to quell the wrath of the storm beasts.

Weather is all around us. We can't escape it, we can't control it, but over time we have learned to understand it. Ancient mythmakers invested the rain and the storm with the same qualities they knew in themselves. Movement meant life. So when early people saw the rains fall and the winds blow, they believed that the rain and the wind acted by force of will. People who knew nothing of science saw a rainbow of color arc across the horizon, and they believed it to be the powerful snake god who had the power to release the rains or to end them.

Early people may not have understood atmospheric science, but in their attempt to make sense of the world they came to understand. Supernatural power inspired both awe and fear, and people had a driving need to make sense of what moved the seasons and changed their world. Early people paid close attention to the weather, and they used myth and storytelling to record their observations. People all over the world created weather myths that document not only the physical movements they witnessed but also the religious beliefs that permeated their culture.

It's amazing to read these myths today and compare the knowledge ancient people learned by simply watching the weather with the knowledge modern meteorologists learned through scientific method. It's also amazing to learn that people in societies all over the world created similar myths because people everywhere have similar needs and fears. Today we see a rainbow and we understand the mechanics of light and color in the air. This makes it difficult to recognize the rainbow snake that arched through the sky centuries ago. We no longer see the storm beast that seemed so threatening, and we no longer see the rain god who sat in the clouds and poured water down to Earth. Yet people everywhere, in the ancient world and today, want to believe in miracles. Read the myths and you'll come to recognize the science of the sky as miracles. Learn the science behind the myths and you'll come to understand how the human desire to unravel the mysteries of the world led people to use their eyes and their minds to build a world of knowledge about our physical universe.

1 .... Clouds and Rain

—————— THE MYTHS OF CLOUDS AND RAIN

Rain seems to fall from the sky like magic. It fertilizes the Earth; it guarantees life. Anyone who can control the rain has miraculous power. In myths, great sky gods controlled the rain. They lived in the clouds, and they poured water from pots or pitchers or simply by force of will. Often in legends powerful kings controlled the rain, and in Africa some people still think their ancestors control it. Myths and legends tell of rain gods herding clouds like cattle. Some tell of drought demons that imprison the cloud cattle in caves and keep them captive until the rain god releases them.

Rainmaking was one of the most common rituals of past times and still exists in some cultures today. Rainmaking combines religion and magic. People who profess to be able to produce rain perform various ceremonies and rituals to stimulate nature to release water. Rain rituals frequently involve both producing the rain and stopping it. In dry lands such as Africa, rainmakers are abundant, and rainmaking myths permeate African culture.

Rainmakers are magicians. They can sing to the clouds to fill them with water and command them to rise to the sky. Sometimes rainmakers use magic charms and medicines to produce rain. Often, they sprinkle or splash water onto the soil or into the air. Rainmakers use numerous techniques in their rituals, such as beating drums to mimic thunder or inducing sparks of fire to mimic lightning. In some cultures, rainmakers mimic the clouds. They use mysterious black potions to turn ordinary clouds into dark heavy clouds laden with water droplets. To stop the rain, rainmakers often produce heat to dry wet objects, such as sand.

In Africa, rainmakers sometimes rely on their ancestors for help in producing or stopping the rain. Ancestral spirits are often protective spirits. In some cultures even today people invest them with the power to protect the Earth from droughts and floods. In Africa and many other lands, kings were invested with mystical powers that they used to protect their people, and often they got these powers from divine ancestors. Commonly, these kings were thought to be earthly counterparts or incarnations of gods or ancestors. The Lovedu have a queen who has more rainmaking power than any of the kings who rule in Africa.

Read the following story of Modjadji, the rain queen of the Lovedu. Modjadji is a real queen, and her rainmaking powers have been the subject of legend for many years. Modjadji rules the Lovedu tribe of South Africa, and she is one of the greatest rainmakers on the continent. The legend of this powerful queen contains beliefs about rain and rainmaking that appear in myths and legends around the world.

"MODJADJI, THE LOVEDU RAIN QUEEN," A MYTH FROM AFRICA

At the foothills of the Drakensberg Mountains in southern Africa there is a tiny kingdom and a powerful queen. The queen has no army and no palace, but she has as much power as if she ruled an empire. The people who live in this small kingdom are called the Lovedu, and the queen is called Modjadji. Modjadji is a great magician. She can transform the clouds and control the rain.

Modjadji gets her power from the spirits of her ancestors, whose magic is so powerful that it can determine the fate of the world. It is by Modjadji's will that the land remains fertile, and it is by Modjadji's will that the Lovedu people survive. Modjadji has learned how to tap the power of her ancestors with the help of a rain doctor and a secret charm. Modjadji's people revere her for her ability to protect them from drought. Her enemies fear her for her ability to withhold the rain and cause devastation to their lands.

This powerful rain queen has ruled the kingdom of the Lovedu for many years. A succession of kings and queens ruled before her, and they, too, tapped into the powers of their ancestors to command nature to behave. Modjadji's people have been rainmakers for centuries, and fire makers, too. They are descended from the great king Mambo who ruled the Karanga people and who lived in a city built of stone in the hills of what is now southern Zimbabwe. Mambo had a potent rain charm and sacred beads, and he kept them hidden for many years so as not to release his secret and not to relinquish his power. King Mambo protected the rain charm for a long time, until one day when his daughter Dzugudini got angry with her father and stole it from him. Dzugudini fled her father's kingdom to a land farther south, and with the charm in her possession she founded the Lovedu tribe.

The founder of the Lovedu was the first of a succession of rain queens who became known as Modjadji. Modjadji I passed on her rainmaking power and her charms to her daughter. This new queen, Modjadji II, became the most powerful rainmaker of all. She needed no army because she could command the clouds. Rain was her greatest gift, and drought was her only weapon. Modjadji II became known to her people as "She Who Does Not Fight." While neighboring tribes fought among themselves, Modjadji shielded her people from harm and ensured them peace and prosperity. Even Shaka Zulu, the violent warrior chief of a tribe much more numerous than the Lovedu, respected Modjadji's power. He left her people to themselves and sent a messenger to visit this powerful queen and ask for her blessings.

Since the time Dzugudini fled from Mambo's kingdom and inherited the rain charm, three Modjadjis have ruled the Lovedu. The current queen has as much power as the queens before her, though she lives a much different life than they lived. People today who learn of a mortal woman who has the ability to control the rain might question the validity of such power. When the Europeans arrived in southern Africa, they questioned the queen's power, and they invaded the queen's territory and changed the way her people lived. The Lovedu people do not question the power of their queen, however. They know that she holds secrets that ordinary people can't begin to understand.

The African people hear many legends about the Lovedu rain queen. They understand that her people believe her to be immortal. This queen is beautiful and mysterious but as inaccessible as a god. Since the founding of the tribe, the rain queen has lived in seclusion in a mist-covered valley deep within the Drakensberg Mountains—untouchable to her people and unbelievable to most everyone else. The Lovedu rarely see their queen because society forbids it. They only see the clouds drifting over her mountains, and they know she is there.

Legends of the Lovedu rain queen tell how Modjadji is as kind as she is powerful. They tell that she cares for her people by giving them rain and ensuring that the years keep moving through the cycle of seasons. Drought is a constant threat in southern Africa, and anyone who lives there understands the devastating circumstances that arise when the drought demons oppose the queen. According to legend, it is with the help of a rain doctor that Modjadji provides relief from the drought. Through magic, the rain doctor determines the causes of drought and then uses powerful medicines to release the rains. In secret rituals deep within the Valley of Devils, the queen and the rain doctor perform their magic. They burn the secret medicines to make black smoke. The black smoke, people say, transforms the clouds. These new clouds are rain clouds—dark gray and heavy with water. They rise to the sky, and then the rains pour down and drench the land.

Little is known about the Lovedu people and less is known about their queen. We do know that the Lovedu once occupied 1,500 square kilometers of southern Africa but that in the 1890s the Boer settlers in southern Africa took away much of this land and gradually Lovedu society began to change. For a long time the Lovedu kept to themselves, and the queen continued to protect her people from outside influence. She blessed her people with rain, and they farmed the land and lived comfortably off the plants and the insects. The Lovedu people used the plants they grew for food and shelter, for clothes and tools, for cooking vessels and for medicines, and for everything they needed in their everyday lives. The Lovedu worked peaceably among themselves, and they relied on no one but themselves for many years. By the grace of Modjadji, the Lovedu had all that they needed, and they were happy. Their queen was a great protector. It is said that once, during a plague in the 1850s, Modjadji made a rain of locusts fall from the sky to nourish her people.

When the Europeans arrived, however, everything changed. Today, the Lovedu live on only a small portion of the land they once had, and they live in the middle of the Republic of South Africa. This small portion of land can no longer sustain the Lovedu people, and Lovedu men and women must go into the towns to work. They must buy food from the Europeans. They must work for money to pay taxes. When the Boers arrived and then the British, these newcomers wanted to unite South Africa. They defeated the Zulu and other neighboring tribes, but Modjadji II surrendered peacefully. She lost much of her land, and her people retreated in secrecy, protected by the Drakensberg Mountains on the south and by the Limpopo River on the north. Lovedu society was in complete opposition to the British Empire and the values they embraced, but they had to live by the rules. The Europeans made the Lovedu pay to live on the land they had worked and farmed for many years.

The current Modjaji rain queen lives in what is now called the Sacred Forest. She lives in a large house made of mud brick and surrounded by giant cyads, plants that covered much of the Earth in the days of the dinosaurs but that remain in only a few places on Earth today. Legend has it that the current rain queen still holds the secret charms and still rules the Lovedu with the power of rain. She conducts her rainmaking ceremony each October under a sacred tree. She beats her sacred drums, she fertilizes the Earth, and her people dance. The current rain queen does not dress like one might expect a queen to dress, and to the outside world she no longer has the mysterious aura of the rain queens before her. The current rain queen wears modern clothes and sits on plastic lawn chairs. She listens to modern music on a boom box outside her home. Reporters visit her, and they take her picture. Yet among her people, the woman is as regal a monarch as she always was. And she is magic. Everyone bows in her presence. No one speaks to her directly. The Lovedu people know that their queen is good and kind and that she protects them. They know that she brings peace and prosperity and rain. Each October the Lovedu people hear the sound of drums coming from the Sacred Forest. They see the mist rise up over the mountains. They feel the rain pour from the sky. By the grace of their queen, the Lovedu people survive. That much will never change.

....

The story of Modjadji the rain queen was created from accounts in *The Realm of the Rain Queen: A Study of the Patterns of Lovedu Society* by E. J. and J. D. Krige (London: Free Press, 1964), and from *Looking for Lovedu*, a book by Ann Jones that describes the author's trip to Lovedu to visit the queen in 2001 (New York: Knopf, 2002).

THE SCIENCE OF CLOUDS AND RAIN

People have always searched for ways to control the rain and weather, and clouds and rain have been the subject of numerous scientific experiments in an attempt to provide relief from droughts. Rainmaking has been around for centuries because people understood that without rain nothing could live. In many myths and legends, the world began with nothing but water. Water had so much power that early mythmakers gave it the power to create the universe. Myths that stress the movement of water through the Earth, sea, and sky revealed an early recognition of what we now know as the hydrologic cycle.

BELIEF: Rain falls from the sky.

Though most people still say that rain falls from the sky, it is actually part of a continuing cycle. Water cycles from Earth, to the atmosphere, back to Earth, through the soil and the seas, and then back up to the atmosphere. This is called the hydrologic cycle. Water from the Earth's surface, from the oceans and lakes and streams and rivers, evaporates into

FIGURE 1.1 · The Water Cycle

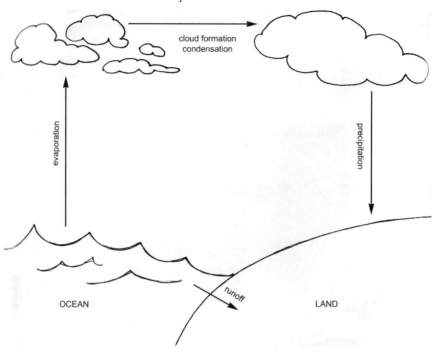

the atmosphere. So does water from the Earth's vegetation. The water vapor rises, condenses into water droplets, and forms clouds. When the water vapor in the clouds condenses and cools, the water droplets form larger droplets. Gravity pulls them downward. When these rain droplets hit the Earth, they sink into the ground and run off into the lakes and rivers and then the oceans. Then the cycle begins again. Take a look at Figure 1.1. Drawing this figure on the board might help to explain the water cycle.

BELIEF: It is possible to transform the clouds and produce rain.

Early people recognized the connection between clouds and rain. Many people also seemed to recognize that not all clouds produce rain. Clouds must contain a combination of water droplets and ice crystals to get thick and heavy enough to produce rain. Clouds that are too high and contain only ice crystals won't produce rain, and clouds that are too low and made up of only water droplets won't produce rain.

Only a small amount of moisture from clouds turns to rain that reaches the ground, so in their efforts to produce rain, scientists have searched for ways to transform more water in clouds to precipitation. In essence, they transform the clouds. This is what Modjadji does. She is called the Transformer of Clouds. The story at the beginning of this chapter explains how Modjadji and the rain doctor use secret medicines to transform ordinary clouds into rain clouds and induce them to rise to the sky and release water. Scientists today use salt crystals. Rainmaking has been around for centuries. Over the years, it has moved out of the realm of magic and into scientific laboratories. Modjadji still makes rain from her mist-covered home in the valley, but rainmakers in other parts of Africa and in other areas of the world engage in this practice, too. Modern-day rainmaking involves a process called cloud seeding. Scientists use salt crystals as their rainmaking medicines. Potassium chloride and sodium chloride both release moisture, and scientists have a way of combining these chemicals and depositing them in clouds.

Cloud seeding has been practiced for more than fifty years. It has been practiced in at least twenty-four countries and in at least ten U.S. states. Groundbreaking experiments have been conducted in dry countries such as South Africa, Thailand, Cuba, and Mexico. In the United States, cloud seeding began in 1946 in a General Electric laboratory in Schenectady, New York. In these first U.S. experiments, researchers used dry ice as a seeding agent. Later experiments used chemicals such as silver iodide, which accelerate the formation of ice crystals.

In ancient times, weather was always sacred knowledge. Early people believed that only gods knew the weather. During times of drought, people who wished for rain had to perform rituals and conduct ceremonies to win the gods' favor and encourage them to produce rain. Rainmakers in different societies gained power because they had connections to the sky powers. They could control the rain by magical or supernatural means that not everyone could understand. In some areas of Africa, rainmakers dressed in elaborate costumes complete with masks, feathers, and animal skins, and they performed dances to encourage the rain to fall.

People began to experiment with more scientific ways of producing rain in times of necessity. Rain was necessary for survival, and people wanted to be able to put a stop to life-threatening droughts. From the 1890s to the 1930s, American rainmakers traveled to drought-stricken areas of the United States promising to make it rain. Some of these rainmakers used sound, some used scents, and others used electricity. These people were considered disreputable, and they gave rainmaking a bad name. At this time, the practice of rainmaking was called pluviculture, a name that derives from the Latin word *pluvial* which means "rain."

One of the best-known pluviculturists was named Charles Mallory Hatfield. In 1904 Hatfield went to Los Angeles and promised the citizens that he could relieve the drought. He released chemicals from two towers, and miraculously rain fell from the sky. Hatfield got $1,000 for making rain in Los Angeles, so he moved on to other states and began charging farmers to make it rain on their fields. Often, he succeeded. Hatfield may or may not have had anything to do with the rain that fell, but he has been credited for making it rain hundreds of times.

BELIEF: Rainmakers protect the people.

Cloud-seeding programs appear to have succeeded on many occasions, and proponents of the process believe they can help farmers protect their crops. Since the invention of agriculture, farming societies have relied on the rain, and they have greatly feared the circumstances should the rain fail to fall. Long ago people who claimed to be sorcerers or magicians practiced rainmaking. They had secret rain charms, and their people looked to them for protection. The Lovedu relied on Modjadji to produce all the rain that they needed.

The powers of reputedly magic rainmakers extended beyond rainmaking, and cloud seeders attempt to extend their control over nature as well. Cloud seeding is undertaken not only to produce rain but also to prevent hailstorms and, in some cases, to disperse fogs.

In Thailand, King Bhumibol Adulyadej takes an active role in helping poor farmers. The king recently received a patent for his rainmaking efforts. He seeds clouds by releasing crystals from aircraft, and to target specific areas he seeds warm and cold clouds at different altitudes. The locals call his project to bring relief Royal Rain.

BELIEF: It is possible to withhold rain.

Cloud seeding can often make rain in smaller amounts by inducing the clouds to produce rain before they build up so much moisture that they release downpours. These efforts are attempts to prevent flooding, but more effective ways include building levees and dams. Levees are artificially raised riverbanks, and they prevent rivers from overflowing and destroying property. Dams are artificial walls that block water. The

United States has about 75,000 dams. Another way to reduce the damage caused by flooding is to establish designated flood zones. Flood zones serve as warnings to builders. If a certain area of the land is susceptible to flooding, governments and builders should heed the warnings and refrain from overbuilding in these areas.

BELIEF: Drought opposes the rain.

The life-giving power of rain is never so evident as in times of drought. In ancient times, people greatly feared drought, and they expressed their fear in myths and legends of drought demons who keep the waters captive and wage battles with rain goddesses to do so. Nature myths are full of storm gods who kill drought demons to release the rain, the most famous of those myths involve heroes who slay dragons. Chapter 5 in this book retells an Indian myth of a battle between a storm god and a drought demon. In this myth, the powerful god Indra controls all the waters of the world. In myths of many lands, dragons control the rain. They often live at the bottom of pools or in the sea where they remain during winter droughts, and they rise up from the water and into the sky in the spring and release the rains when they fight battles in the clouds.

BELIEF: Rain is a blessing and is necessary for survival.

Water sustains life, and, in certain areas of the world, rainfall is inadequate. People long ago knew this, and they relied on rainmakers like Modjadji to relieve drought and release the rains from the sky. Rain sinks into the soil, and it nourishes the plants and animals. This is certainly a blessing and essential for the continuation of life. If rain falls more than an inch or two in an hour, however, it does not sink into the soil: it runs off the soil and causes flooding.

Flood stories arose in almost every culture in almost every area of the world because everyone was familiar with the devastating effects of too much water. You may have heard of a flash flood warning. This is caused by a sudden rainstorm that comes quickly and often with little or no warning. Flooding can occur unexpectedly, and people can drown because they are unprepared. Floods can occur for reasons other than rain, though. Floods can occur when excessive wind causes the ocean waves to surge onto the shore. Floods can occur when ice melts from the mountaintops or when heavy rains cause landslides and mudslides. Landslides and mudslides push debris from the mountains to the ground that blocks rivers and lakes. This is a frequent problem in the hills of California, largely due to excessive excavation and construction. As builders develop homes and office complexes on the mountaintops, they must cut into the hills. This causes the land to become unstable, and when it rains the land breaks up and tumbles to the ground.

— TOPICS FOR DISCUSSION AND PROJECTS

TOPIC 1. Myths, Legends, and History

Much of what we know of the Lovedu was never written down but was passed down orally. This is true of many societies in Africa. With oral history, myth, legend, and history are so intricately entwined that it is often difficult to distinguish fact from fiction.

PROJECT IDEA

Take time to discuss the differences between fact and fiction, history and legend. Work with students to identify the characteristics of each genre. Define fiction and nonfiction and give examples of each. Tell students that fiction is based on imagination and nonfiction is based on fact. Ask students to identify examples of fiction and nonfiction. Examples of fiction are novels, short stories, poems, and plays. Examples of nonfiction are biographies, news stories, histories, and memoirs. Tell students that while these examples can be clearly classified as fiction or nonfiction, myth, legend, and history often contain elements of both. Folktales and fables are other genres you might want to discuss. Folktales and fables also have elements in common with myths, legends, and histories.

You might want to place the information in a Venn diagram such as in Figure 1.2. When the diagram is finished, students will see how the three genres overlap. Discuss with students how myth, legend, and history are connected and confused.

FIGURE 1.2 • Venn Diagram

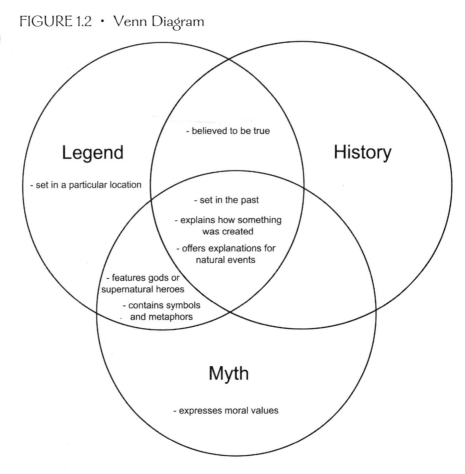

SUGGESTED READING

Mythology Web. *http://www.mythology.com/*.

> Contains myths, legends, and folklore from many areas of the world. Compares stories with similar themes and has informative articles about these genres.

Rosenberg, Donna. *Folklore, Myth, and Legend: A World Perspective Teacher's ed.* Lincolnwood, IL: NTC Publishing, 1997.

> Discusses the differences between genres and includes 120 tales from many lands, including Africa. Includes historical and cultural background on the various motifs and mythological subjects.

TOPIC 2. Mysterious Characters

Part of Modjadji's appeal was her mystique. Modjadji's womanhood added to her mystique and so did her seclusion in the mist-covered mountains. Most people never saw Modjadji but only heard stories about her. The stories were intriguing enough to fire the imagination, and legends and stories have surrounded the powerful rain queen since the founding of the Lovedu tribe.

PROJECT IDEA

Tell students to think about what makes someone mysterious. We already said that Modjadji was a powerful woman and that she remained hidden in mist-covered mountains. She also had a rain charm and a power no one understood: the supernatural power to control the rain. Discuss mysterious characters students might recognize from fiction and theater, such as Harry Potter or the Wizard of Oz. Have students identify other mysterious characters they are familiar with and discuss what makes each of these characters mysterious. Then have students create their own mysterious character and write a character analysis. Their character should have some kind of power, some kind of attribute or weapon, some kind of mysterious domain, and some kind of lifestyle that adds to his or her mystique. You may want to use a chart like the following to help students determine what to include in their writing.

Character	Power	Attribute	Domain	Lifestyle
Modjadji	Rainmaking	Rain charm	Mist-covered mountains	Remains hidden
Harry Potter				
Wizard of Oz				

Legends and folktales from Africa and other lands were often told as explanations of events that people witnessed in the natural world. The tales of explanation have been referred to as why stories, and why stories are particularly present in African oral tradition.

PROJECT IDEA

Have students read legends from African oral tradition that were told to explain an occurrence in the natural world. Then have students write a legend of their own that explains something. Their legends can explain why the clouds are puffy, for example, or why the cat has whiskers. The books in the Suggested Reading section contain selections of legends and folktales and will help students gain an understanding of the structure and composition of why stories.

SUGGESTED READING

Ardagh, Philip. *African Myths and Legends*. Chicago: World Book, 2002.

> Contains a selection of myths and legends from sub-Saharan Africa. Includes a list of resources and a glossary.

Arnott, Kathleen. *Tales from Africa*. New York: Oxford University Press, 2000.

> Contains many tales that give explanations for occurrences in the natural world.

Young, Richard, and Judy Dockrey Young. *African-American Folktales for Young Readers*. Little Rock, AR: August House Publishers, 1997.

> Contains folktales from African oral tradition that have been told by African storytellers in the United States.

TOPIC 4. Water Conservation

Water covers 71 percent of the Earth's surface. Of the Earth's water, 99.7 percent is in the oceans, the soil, glaciers, and the atmosphere. The remaining 0.03 percent is available for human use, though much of it is inaccessible. Human beings consume about 398 billion gallons of water a day, and all of it comes from surface water and groundwater. Surface water refers to bodies of water that are visible on the Earth's surface, such as oceans, rivers, lakes, and streams. Groundwater refers to water that is in the soil or in underground aquifers.

Most water humans use comes from rivers, but in some areas it comes from aquifers. Aquifers are like underground lakes. In the United States, there are very large aquifers in South Dakota, Nebraska, Kansas, Oklahoma, and Texas. People drill wells into aquifers to bring water to the surface, and then they use that water for crops and for drinking. It is extremely important to protect these underground water sources and to take care not to pollute them.

PROJECT IDEA

Have students brainstorm ways they can conserve water. Then make a mural or a bulletin board to post in the school hallway or design a pamphlet that students can take home to their parents. Stress to students the importance of conserving water. Write different ways to conserve water on the board, and then have students include this information on the bulletin board or on their take-home pamphlets. Following is a list of some of the ways people can conserve water. Discuss these ways in class and have students add their own ideas to the list. The Web sites listed in the Suggested Reading section contain information on various ways to save water.

What You Should Do Every Day

Do not leave the water running while you brush your teeth.

Take showers instead of baths.

Water the lawn only at night or in the morning.

Run the dishwasher and washing machine only for full loads.

Use the lowest water level possible for washing machine loads.

Recycle water. Reuse dishwater to wash the car and reuse bathwater to water plants.

Do not leave the hose running when you wash your car; use water from a bucket instead.

Instead of rinsing dishes before washing them, just scrape them.

What You Should Do During Extended Dry Periods and Droughts

Take shorter showers.

Do not water the lawn more than necessary.

Do not wash your car more than necessary.

SUGGESTED READING

American Water and Energy Savers. "Save Water 49 Ways." *http://www.americanwater.com/49ways.htm*.

> Lists forty-nine ways to save water.

The Groundwater Foundation. *http://www.groundwater.org/*.

> The Kids' Corner section has lots of information on underground water sources and instructions for building an edible aquifer. Students might want to build an edible aquifer as a class activity.

National Drought Mitigation Center. "Drought Mitigation." *http://www.drought.unl.edu/kids/reduce/mitigation.htm*.

> Contains information on droughts and water conservation. Has ideas for ways to conserve water in day-to-day living.

New Jersey AWWA. "Kids Water Zone." *http://www.njawwa.org/kidsweb/htm*.

> Contains lots of information on the water cycle and the use of water. Includes tips for conserving water in the home and in life, and includes ideas for activities and projects.

TOPIC 5. Types of Clouds

The story of Modjadji makes reference to different kinds of clouds. Modjadji the rain queen transformed one kind of cloud to another. There are numerous types of clouds that help determine what effects we feel on Earth. Cumulous clouds are thick and flat. Stratus clouds are flat and layered. Cirrus clouds look wispy and are made of ice crystals. Nimbus clouds produce precipitation.

PROJECT IDEA

Make a cloud chart for the classroom that shows the different types of clouds and what they do. The Suggested Reading section lists Web sites that have both information about and photographs of the types of clouds.

SUGGESTED READING

Boatsafe for Kids. "How to Be a Storm Spotter." *http://www.boatsafe.com/kids/weather1. htm.*

> Clear, concise definitions and descriptions of the different types of clouds.

Plymouth State University Meteorology Program. "Cloud Boutique." *http://vortex. plymouth.edu/clouds.html.*

> Gives explanations and provides detailed pictures of cloud formations.

University of Illinois Department of Atmospheric Sciences. "Clouds and Precipitation." *http://ww2010.atmos.uiuc.edu/(Gh)/guides/mtr/cld/home.rxml.*

> Contains classroom activities, pictures, and information about the different types of clouds. Links to an online guide to meteorology that contains more information about and activites on clouds for use in grades 5–8.

TOPIC 6. African Tribal Society

The Lovedu belong to a larger group of Africans called the Bantu. The Lovedu live in huts with roofs made of straw, each hut with its own field and granary. The Lovedu are agriculturists, and they rely on plants of many kinds in many aspects of their lives. These people were self-sufficient, but they relied on their land and their queen to provide for them. They had a special bond with nature and an extremely strong sense of family. Children were brought up by their grandmothers. Young children herded cattle by the age of five. In Lovedu society, children learned by watching their elders. Children were praised and not criticized because the Lovedu upheld a philosophy of peace in all areas of life. Because Lovedu children grew up peacefully, they did not fight like European children often did. They were taught the value of peace, generosity, and unselfishness from the day they were born.

Miraculously, Lovedu society remained unchanged by the Europeans until after World War II. They lost much of their land, and the little land they did retain was not enough to feed their people. They had to buy maize from the very farmers who had stolen their fields. The Europeans banned the cutting of trees, so the Lovedu could no longer build their homes or make their tools like they always had. They had to buy tools and cooking vessels from the Europeans. By the 1970s, Lovedu women had to work for the Europeans to make money.

PROJECT IDEA

Have students write a report on the traditions and customs of one of the African tribes. In addition to using general encyclopedias available in the reference section of the school library, students might want to use the sources in the Suggested Reading section.

SUGGESTED READING

Haskins, James, and Joann Biondi. *From Afar to Zulu: A Dictionary of African Cultures.* New York: Walker and Co., 1998.

Covers Africa's thirty-eight best-known cultures. Includes history, geography, and culture.

Murray, Jocelyn, ed. *Cultural Atlas of Africa.* New York: Checkmark Books, 1998.

Explains the culture and history of African people throughout the continent. For ages 9–12.

Musgrove, Margaret. *Ashanti to Zulu: African Tradition.* New York: Dial Press, 1976.

Describes the customs and traditions of twenty-six African tribes.

—— SUGGESTED READING FOR TEACHERS

Dennis, Jerry. *It's Raining Frogs and Fishes: Four Seasons of Natural Phenomena and Oddities of the Sky*. New York: HarperCollins, 1992.

> Explains the beliefs and trivia surrounding the world weather.

Donnan, John A., and Marcia Connan. *Rain Dance to Research*. New York: David McKay, 1977.

> Discusses rainmaking in history, culture, and scientific studies.

How the Weather Works. *http://www.weatherworks.com/*.

> An educational site on the weather for teachers and students. Has workshops and multidisciplinary study units. Also has information on purchasing educational products.

Jones, Ann. *Looking for Lovedu: Days and Nights in Africa*. New York: Knopf, 2002.

> Travel writer recounts her adventures in Africa and her search for Modjadji, the Lovedu rain queen.

Krige, E. J., and J. D. Krige. *The Realm of the Rain Queen: A Study of the Patterns of Lovedu Society*. London: Free Press, 1964.

> The only anthroplogical account of the Lovedu Tribe.

National Groundwater Association. "Information on Earth's Water." *http://www.ngwa.org/educator/lessonplans/earthwater.htm*.

> Discusses the waters of the world, floods, and flood control and provides lesson plans for teachers.

Oracle Education Foundation and ThinkQuest. Africa: Cradle of Civilization. "Modjadji, the Rain Queen of the Lovedu." *http://library.thinkquest.org/C002739/AfricaSite/LMSouthModjadji.htm*.

> Explains the history and legends surrounding the Lovedu rain queen.

St. Francis Drake High School. "Floods and Droughts." *http://drake.marin.k12.ca.us/stuwork/rockwater/Floods%20and%20Droughts/mainpage.html*.

> Comprehensive student report on floods and droughts. Contains a great resource list.

2 ········· Thunder and Lightning

THE MYTHS OF THUNDER AND LIGHTNING

In myths, thunder and lightning were evidence of sky power. They seemed to come from the sky, and they appeared to be used as weapons that sky gods used to strike the Earth when they got angry with human beings. In some myths of thunder and lightning, these forces were evil, but in other myths they were good and kind. Long ago, agricultural people worshipped the thunder and lightning because they understood their ability to restore fertility to the Earth after long winters and devastating droughts.

People everywhere understood the dual nature of thunder and lightning. They could bring rain when it was needed, or they could destroy the Earth with their strength and might. Ancient people did not understand what caused the thunder and lightning, but they knew that often they came in the spring. They also knew that birds came in the spring. You might see how it was easy for people who could not explain the weather scientifically to associate thunder and lightning with a gigantic bird. When the thunder boomed, it sounded like the flapping of giant wings. When the lightning flashed, it looked like the blinking of giant eyes.

Myths and legends include many metaphors for thunder and lightning, and many stories describe thunder as the sound of a bird. In some stories, thunder was the bellowing of a ram or the howling of a dog. In other stories, it was the hissing of worms or snakes, and in still others it was the sound of fighting dragons or the rumbling of chariots in the sky. In legends, lightning was often considered fire in the sky—fire that flickered like magic or flashed from the bright flaming eyes of serpents or snakes or birds that brought rain and storms. It was a widespread belief in the ancient world that birds controlled the weather. This belief appeared in myths from the Americas, China, Africa, and many other lands.

Read the following story, which was adapted from a legend the Passamaquoddy told about Thunderbird. The Passamaquoddy is a tribe of Native Americans that made their home long ago in parts of what is now the state of Maine and the province of New Brunswick, Canada.

"THUNDERBIRD," A LEGEND FROM NORTH AMERICA

It was late one afternoon when suddenly the wind rushed over the plains and dark clouds covered the sky. Soketao had been walking in the forest, as he always did before sunset, but suddenly he stopped and looked at the sky. It was time to call the children together, he thought. He would build a fire in his wigwam, and the children would gather around it, their legs crossed and their eyes full of wonder. Soketao would talk to them, and he would tell him them about Thunderbird. He would tell them to watch for Thunderbird, and he would help them understand.

Things are not always black and white or good or bad; that would be Soketao's lesson. Thunderbird was powerful, and the children had better learn that they should take cover when the bird approaches. But just as important, Soketao knew, the children must learn proper respect for Thunderbird. They must understand his power and know his ability to renew the Earth just as surely as they knew his ability to destroy it.

Soketao walked quickly as the sky grew darker and darker. He tapped his walking stick on the ground and held his hand over his brow to shield his eyes from the wind. Soketao could tell that Thunderbird was fast approaching. Soketao could hear the rumbling in the distance—he always could, long before he saw Thunderbird's eyes flash in the darkness and long before he felt the rains pour from the sky.

By the time Soketao reached his home it was dark. Several of the children had already gathered outside the old man's wigwam, waiting patiently for him to return and welcome them into his home. For as long as anyone remembered, Soketao had been telling stories to the children. He had told them to these children, and to their parents, and even to their parents' parents. Soketao was a very old man, and he had lived long enough to tell stories that had been told for centuries before anyone today had ever been born. These were stories that Soketao knew well, and these were stories that Soketao loved to tell, for he knew that it was important for children today to understand the powers of the world just as children long ago understood them. Thunderbird would come every year; that was certain. It was important to recognize the signs.

"Gather 'round, children," the old man said. The children entered the wigwam and took their places around the fire pit in the middle of the room. More children came, and then some others. Soon, the wigwam was full of chattering voices. Soketao lit the fire and took his place in the circle.

"Come in and hear of Thunderbird," he said, motioning to the children who were standing at the door. When all was silent, he began his story.

"A long time ago," Soketao began, "two young boys set out to find the origin of thunder. They left their home in the forest, and they traveled far to the north and high into the mountains. Soon, they came to a magic mountain range, with snow-covered peaks that jutted high into the sky."

Soketao looked around the room. The children sat still and quiet. Some of them huddled together.

"The two boys had never seen a magic mountain range before," Soketao continued, looking around the room. "But they knew it was magic because the mountains split apart and then moved back together, right before the boys' eyes. The boys watched in amazement as the mountains separated and moved together and then separated and moved together again. Each time the mountains moved, they left a large cleft—just for a moment. The boys strained their eyes to look inside the cleft, and they saw right before them what appeared to be an entranceway to a mysterious land.

" 'I think we have found the origin of thunder,' one boy said to the other. 'I think the thunder comes from inside the mountains! The next time the mountains open, I will jump into the cleft quickly before it closes. You follow me, and we'll see what we can find.'

"The other boy nodded his head in agreement, and the two of them waited silently until the mountains opened again. Then, quick as a wink, one boy jumped and then the other. The first boy landed with a thump on a large plain inside the mountain. The second boy never made it through. The mountains closed on him before he could jump inside."

The youngest children around the fire pit always gasped at this part of the story, and their eyes widened in fear. They were afraid for the young boy who had failed to jump into the mountain, but they were also afraid for the other young boy who found himself in a mysterious land all alone.

"What happened to the boy who didn't make it through?" someone always asked. And then someone else always told the inquisitive young child to be quiet and let Soketao finish his story.

"The boy who landed inside the mountain was now a lone Indian. We will call him the lone Indian. He knew that something had happened to his friend, but he was transfixed by the sights and sounds inside the mountains. The lone Indian was amazed to see that this strange new land looked very much like the land he had left behind. He looked around and tried to decide which way to go. Then suddenly, from the north, he heard sounds. It sounded like children playing ball.

"The lone Indian was inquisitive and walked forward. He came to the edge of a cliff and he looked down into a valley where he spotted what looked like a large group of wigwams at the edge of the sky. The lone Indian walked carefully down the edge of the cliff and toward the wigwams. Then suddenly he stopped. There were children running and playing outside the wigwam; he could see that for certain now.

"The lone Indian was afraid to get too near the children for he did not know if they would welcome him or if they belonged to a hostile tribe. He sat down behind a tree and watched them from a distance. The children never noticed the boy, and eventually they finished playing and disappeared into their wigwams.

"The lone Indian waited for a long time and considered whether to move or to stay right where he was. The sky began to grow dark, and gray clouds floated overhead. He was just about to walk farther toward the village, realizing that perhaps a storm was approaching and that soon he would need to seek shelter. But before he could move, he saw the children emerge from their wigwams. They wore large wings strapped to their backs, and they carried bows and arrows that they had not had when they went inside. The lone Indian watched in amazement as the sky grew dark and loud thunderous booms clapped in the sky overhead.

" 'This must be the home of the thunder,' he whispered to himself, hoping that his friend had somehow managed to jump inside the mountain and would soon be there to accompany him. Before the lone Indian had time to consider this, the children spread their enormous wings and rose up from the ground.

"Thunder boomed through the clouds as these bird children flew overhead and toward the south," Soketao explained to the children, now wide-eyed with wonder. "Rain poured from the heavens, and the lone Indian ducked into the bushes to protect himself. A bolt of lightning split the tree right behind him and he ran. He ran as fast as he could toward the wigwams. The sky had grown so dark that he could barely see, and before he knew it he ran right into an old man. The man had his arms crossed in front of him. The lone Indian fell backward on the ground, scared and soaking wet.

" 'Who are you and why do you come here?' the man asked.

"The lone Indian believed he should answer the man kindly, and he did. 'I came to find the origin of thunder,' he explained.

"The man stood silently for a while before he spoke. 'Come with me,' he said.

"The lone Indian was frightened, but he followed the man into a cave. It was dark and cold in the cave, but it was dry. The boy could still hear the sounds of thunder booming outside, and when he looked behind him, he could see streaks of lightning flash across the opening of the cave."

Soketao paused for a few moments and remained perfectly quiet.

"Something strange and mysterious happened inside the cave," Soketao finally said, looking around the room. The children listened intently. No one moved a muscle.

"As the lone Indian and the man continued walking," Soketao said, "the lone Indian forgot who he was and why he had come. He had no sense of time or place. The man and the lone Indian walked deeper and deeper into the cave.

"Thunder roared and a large bolt of lightning flashed. Then, the lone Indian emerged from the cave. He was not the same boy he was when he went in the cave, though. He was a strange boy with the body of a large bird and gigantic wings made of red and gold. He carried bows and arrows and he looked toward the sky.

" 'You now have the swiftness of lightning and the power to renew the earth,' the old man told the boy. 'Do not fly too close to the trees or you could be killed,' the man warned. 'And beware of Wochowsen. Wochowsen is a bird much like yourself, and he too has great power. But Wochowsen's power is evil. He will harm you, and he will harm the world.' "

"Did the lone Indian become Thunderbird?" a young girl asked, huddled in the corner of Soketao's wigwam.

The old man nodded. "Thunderbird is an Indian, and if you respect his power and learn to recognize it, he will not hurt you. He can water the Earth, and he can make the plants grow. But remember, too, the evil Wochowsen. He can split trees, and he can flood our villages. We must learn to recognize his warnings, too, and we must respect his power. He can make the sky grow dark. He can make the wind blow. Thunderbird will fight Wochowsen, you can be sure of that, and Thunderbird is very strong. Thunderbird can make lightning with the blink of an eye, and he can make thunder with a flap of his wings. When he shakes his feathers, the rain falls. We must never underestimate the Thunderbird's strength. Thunderbirds fly from the north and bring rain and thunder throughout the year, but the greatest Thunderbird of all comes every spring. He renews our world after the long cold winter, and he takes care of us."

. . . .

"Thunderbird" was adapted and expanded from the Passamaquoddy legend as it appears on various Web sites.

THE SCIENCE OF THUNDER AND LIGHTNING

This Passamaquoddy legend tells how the young boy passed through the cleft in the mountains and became the great and powerful Thunderbird. In this legend, Thunderbird was good, and he always looked out for the people. The legend also explains that Thunderbird is very powerful. Read the following beliefs that appeared in the Passamaquoddy legend of Thunderbird and compare these beliefs to the science that emerged years later.

BELIEF: Thunder and lightning come from the sky.

Thunder and lightning appear to come from the sky, but today we know that these forces are not celestial but atmospheric. They originate in clouds. Lightning is caused by static electricity that forms from air currents moving inside clouds. Thunder is simply the noise lightning makes. Myths and legends often use the term *thunderbolts*, which mythological sky gods hurl down to Earth when they're angry. In myths the term *thunderbolt* is used to mean thunder and lightning that occur together.

Lightning is caused by static electricity. Water, like all forms of matter, is made up of atoms, and atoms contain negatively charged particles called electrons. Lightning is an electrical spark that occurs when electrons inside water droplets move very suddenly inside storm clouds. During a thunderstorm, water droplets form inside clouds, and strong winds blow the water droplets to the top of the cloud. As they rise, they freeze into slivers of ice. Some of those slivers grow larger and turn into hail. When the hail gets too heavy for the cloud to hold, the hail drops through the cloud and crashes with the rising ice slivers. During these collisions, electrons move from the ice slivers to the hail. Because electrons have a negative charge, these collisions give the hail a negative charge and leave the ice slivers with a positive charge. The hail moves to the bottom of the cloud and the ice slivers remain at the top. Positively charged particles attract negatively charged particles like magnets. Inside storm clouds, the positively charged atoms at the top of the cloud pull on the negatively charged particles at the bottom of the cloud, and the electrons shoot upward. This causes an electrical spark we see as lightning. Sometimes the lightning stays inside the cloud. Sometimes it shoots from one cloud to another. The kind of lightning we see most often shoots between clouds and to the ground.

One interesting fact about lightning is that, though it appears to come from the sky, by the time we see the flash it is actually moving upward from the ground. The positive charges in the atoms on the ground pull the negative charges from the electrons at the bottom of the cloud downward. When the electrons reach the ground, they light up. The light moves upward as more electrons shoot down to the ground. This is what we see when we see a lightning flash—this upward movement of light. Sometimes we see one flash after another. When this happens, the flashes come so close together that the lightning we see appears to flicker.

BELIEF: Thunder and lightning are caused by movement.

In the myth of Thunderbird, thunder is caused by the flapping of wings and lightning by the blinking of giant eyes. This reflects a belief that these two forces are caused by movement in the sky. We have already learned how lightning is caused by the movement of electrons. Thunder occurs from this same movement. As lightning flashes upward from the ground, it heats up the air. This makes the air expand and explode. First we see the flash, and then we hear the explosion. The flash and the explosion occur simultaneously, but we see the flash first because light travels faster than sound. Usually we hear thunder about five seconds after we see the lightning streak. When the lightning is nearby we hear the thunder as a loud clap. When the lightning is farther away we hear the thunder as a rumble. In this case we're either hearing explosions from different parts of the lightning streak, or we're hearing echoes of thunder that bounce off of mountains. When the lightning is too far away, we can still see the flash, but we don't hear the thunder.

BELIEF: Thunder and lightning come hand in hand
with rain.

Believe it or not, lightning hits the ground about 100 times a second, or approximately 20 million times a year in the United States alone. Sometimes it's accompanied by thunder, sometimes by thunder and rain, and sometimes it flashes across dry skies with no rain or thunder.

Streak lightning is the most common form of lightning. That's the kind that produces the characteristic zigzag shape. Forked lightning is streak lightning that splits into two branches. Ribbon lightning is streak lightning that looks like horizontal bands because it is blown sideways by the wind. Bead lightning is broken into small segments, and ball lightning looks like balls of fire the size of fruit.

Not all lightning is dangerous. Sometimes it's so far away it can never get anywhere near us. Sheet lightning lights up entire clouds. Sheet lightning is actually a reflection of lightning flashing behind the clouds. Heat lightning occurs when there are no thunderclouds in the sky, often occurring on hot summer nights. Heat lightning is very far away, possibly even a reflection of lightning that's flashing below the horizon. Because heat lightning is so far away, we don't hear any thunder.

In the myth we just read, Thunderbird lives in the far north in a land of blowing wind and driving rain. In legend, no one could see Thunderbird because he flew behind the dark clouds. People on the Earth could only see his eyes flash and hear his wings flap. They never saw the bird itself, which made legendary heroes everywhere go in search of the bird's home and where thunder originated.

BELIEF: Thunderstorms come every spring.

Most people are familiar with the saying "April showers bring May flowers." In many parts of the world it typically rains in the spring, and oftentimes the rain is accompanied by lightning and thunder. In storm myths around the world, thunder, lightning, and rain were all attributed to Thunderbird. This creature flapped his wings to cause thunder, blinked his eyes to cause lightning, and shook his feathers to cause rain, and people could count on Thunderbird to make an appearance in the sky every spring. Long ago people all over the world recognized thunder and lightning as seasonal powers.

Thunderstorms can occur any time of year, but they typically occur in the spring and summer. This is because most lightning originates in clouds that form at this time of year. Most lightning comes from cumulonimbus clouds or thunderheads, which form when the air gets warm and moist. In much of the world this is when large air masses from the tropics move toward the poles, and when they do they move through the middle latitudes and bring warmth and moisture. In the United States, thunderstorms are most common from May through August and in states along the Gulf of Mexico. In the tropics, where the air is nearly always warm and moist, thunderstorms are common year-round. Near the poles, thunderstorms are rare. This is because in the polar regions the air rarely gets warm or moist enough for cumulonimbus clouds to form.

BELIEF: Thunderstorms warn of their arrival.

Cumulonimbus clouds don't start as cumulonimbus clouds, however; they start as cumulous clouds. Cumulous clouds are white and puffy and can be incredibly beautiful. Sometimes they reflect colors from the sun and glow in the sky. When they grow into thunderheads, though, they get darker and heavier and can look absolutely terrifying. Cumulous clouds turn to storm clouds as warm air rises and hits cooler air. Then water droplets in the clouds turn to ice slivers and hail. As warm air rises, it expands, and as more warm air rises, the clouds blow up like balloons. The clouds get larger and heavier until strong winds begin to blow. When the winds blow, they blow through the clouds and cause the hail and ice slivers to collide. As explained earlier, this is when lightning occurs.

BELIEF: Thunder and lightning are powerful forces.

People always recognized thunder and lightning as powerful forces. They had the power to renew the Earth, but they also had the power to destroy it. In some myths, Thunderbird was good and kind, and in other myths Thunderbird was evil and extremely dangerous. Still, some native tribes in Africa worshipped the lightning bird as well as the Thunderbird. Some of the Zulu people consider the lightning a large bird, shiny red like fire, with large feathers that spread out behind it like the feathers of a peacock.

Thunderstorms have a tremendous amount of energy, which is released as rain, hail, wind, thunder, and lightning. Excessive rain has the power to flood large areas of the land; hail causes tremendous damage to property, crops, and people; and excessive wind can take the form of hurricanes and tornadoes. Thunder is simply sound, so thunder alone is not dangerous. We know that thunder comes from lightning, however, and lightning is the most dangerous aspect of a thunderstorm. According to Safeside, an organization that promotes weather safety, a lightning bolt can be as hot as 50,000 degrees farenheit. That's five times as hot as the Sun. Each streak is a gigantic electrical spark, and when that spark shoots down from the clouds, it doesn't always hit the ground. Usually, lightning hits the highest object it can find. It can cause forest fires, and it can damage powerlines, and sometimes it can cause blackouts in entire cities. When lightning hits trees, it can split them apart; when lightning hits tall buildings, it can cause them to burst into flames. In myths thunderbirds can split trees with their powerful claws.

— TOPICS FOR DISCUSSION AND PROJECTS

TOPIC 1. Birds and Storms

People told stories about Thunderbird long ago to explain how storms and rain came to Earth. Birds traditionally symbolized celestial power. Their ability to fly enabled them to move from Earth to the sky, and their wings were powerful enough to carry them across the heavens. Myths all over the world connected birds to the sky powers. Birds appeared as solar symbols, and people all over the world believed that birds could tap into the power of the Sun, the clouds, and the mighty storms. Native Americans all over the country worshipped birds as wind gods. Most every culture in Native America believed that birds of one kind or another flapped their wings and caused the winds and storms.

PROJECT IDEA

Talk to students about Thunderbird myths from different areas. Divide the class into groups and assign each group a different Native American culture. Have each group choose one myth from their culture and relate that myth to the class. You might want to divide the country into areas to be sure that myths from all areas of the country are represented. The sources in the Suggested Reading section contain legends of Thunderbird and other legends that explain the weather.

SUGGESTED READING

Ewebtribe. "Native American Culture: Stories/Legends." *http://www.ewebtribe.com/ NACulture/stories.htm.*

> Contains legends of the thunderbird and other legends from Native American cultures.

The Natural World. "Thunderbird and Lightning." *http://www.snowwowl.com/ rlthunderbird.html.*

> Contains information on thunderbird legends across the country, information on the connection between thunderbirds and lightning, and pictures of the different kinds of lightning.

"Thunderbirds: Legend from the Pacific Northwest." *http://www.angelfire.com/realm/ bodhisattva/thunderbird.html.*

> Contains information on thunderbird legends and links to legends of the thunderbird throughout Native American cultures.

Representatives from Safeside claim that almost all lightning-related injuries could be prevented if people took cover during lightning storms. Stress to students the importance of seeking shelter indoors. Then explain some of the ways students can prevent being struck by lightning.

PROJECT IDEA

Have students create posters about lightning safety. Have them divide their posters into two columns, one labeled "safe" and one labeled "unsafe." Have students write lists for each column and illustrate each item on their lists. They might want to illustrate their posters by cutting pictures from magazines or creating pictures out of colored construction paper. Some of the things they might include on their safe list would be a storm cellar, a basement, and a house. Some of the things they might include on their unsafe list would be trees, tall buildings, metal objects, wire fences, metal screen doors, water, electrical appliances, and open fields. A good way to make this kind of a poster is to use a heavy material for the poster board and fold the poster board in half. This way the posters can stand up, and students can display their work on their tables or desktops.

SUGGESTED READING

Lightning Protection Association. *http://www.alrci.com/info/*.

> Lists numerous facts and misconceptions about lightning.

Mandell, Muriel. *Simple Weather Experiments with Everyday Materials*. New York: Sterling Publishing, 1990.

> Contains dozens of easy experiments for kids ages 9–12. Has supply lists, complete instructions, and clear explanations about what happens in each experiment and why.

Safeside. "Weather Safety." *http://www.weather.com/safeside/lightning/index.html*.

> Contains information about lightning safety and disaster preparedness plans. Includes common misconceptions about lightning.

TOPIC 3. Native American Storytelling

Soketao is from a tribe of people called the Passamaquoddy, and he tells the children that the Passamaquoddy people told stories to explain things they did not understand.

PROJECT IDEA

Ask students what comes to mind when they think of thunder. If they did not know what thunder was, what would it be? What does it sound like? What noise does it make? Thinking about thunder in this way will help students create images of thunder and incorporate those images in a thunder legend of their own. Write the following list of thunder metaphors on the board. Then have students create their own thunder myths using one of the thunder metaphors or have them create a thunder myth based on what they think thunder might be.

Lightning and Thunder Metaphors

Rumbling of the sky gods' chariots

Voice of the Great Spirit

Bellowing of a ram

Howling of a dog

Hissing of worms

Flapping of Thunderbird's wings

Man kicking things

SUGGESTED READING

Andrews, Tamra. *Legends of the Earth, Sea, and Sky: An Encyclopedia of Nature Myths*. Santa Barbara, CA: ABC-CLIO, 1998.

> Contains lengthy articles on the myths and legends of thunder, lightning, and other aspects of storms.

Caduto, Michael, and Joseph Bruchac. *Keepers of the Earth: Native American Stories and Environmental Activities for Children*. Golden, CO: Fulcrum Press, 1997.

> Contains stories of the thunder and other atmospheric phenomena and activities for students ages 9–12.

TOPIC 4. Ben Franklin's Kite Experiment

It is commonly believed that Ben Franklin proved that lightning was electrical by flying a kite in a thunderstorm. Whether or not this really happened, such an experiment could have killed Franklin, and quite likely it would kill anyone who tried to duplicate such a foolish act.

PROJECT IDEA

The following passage explains how sources commonly recount Franklin's kite experiment. Following the passage are directions for writing a critical essay about the passage. Make copies of this passage and distribute the copies to students. The directions preceding the passage ask students to write an essay to explain if and why they think this experiment is dangerous. The directions also ask students to explain how such an experiment could be done safely. After students turn in their essays, explain to them the dangers of such an experiment. Explain why most scholars today believe that the famous kite experiment is strictly legend. The sources in the Suggested Reading section explain what scholars think of Franklin's kite experiment and provide information on Franklin's other experiments with electricity.

SUGGESTED READING

Birch, Beverley. *Benjamin Franklin's Adventures with Electricity*. Hauppauge, NY: Barron's Educational Series, 1996.

Discusses Franklin's experiments and achievements in electricity.

USHistory.org. "Franklin and His Electric Kite: An Experiment That Took the World by Storm." *http://www.ushistory.org/franklin/kite/*.

Recounts Ben Franklin's kite experiment that allegedly took place in Philadelphia.

Directions

Read the passage below about Ben Franklin's famous kite experiment. Write an essay to explain what you think of this story. Explain if you think this experiment is dangerous and why. Then explain why you think it is true or untrue. How could this experiment be conducted safely?

Ben Franklin's Kite Experiment: Truth or Fiction?

They say that Ben Franklin flew a kite one day in Philadelphia and drew lightning from a cloud. They say that Ben Franklin wanted to prove that lightning was simply a large electric spark, and to prove it he tied a key to a kite, flew the kite in a thunderstorm, and waited for lightning to strike the key.

Anyone who has read about Ben Franklin knows that he was fascinated with electricity and that he conducted many experiments that led to significant discoveries in the field of meteorology. Franklin's kite experiment was one of his most famous—and his most dangerous. It is widely reported that Franklin built the kite he flew that day with the help of his son, William. The two of them attached a sharp metal wire to the top of a kite, and they tied a silk ribbon to the end of the kite string. Then they tied a metal key to the end of the silk ribbon.

Franklin was excited about his kite, and he waited for a stormy day and then took it outside and flew it in the middle of a thunderstorm. Just as he had hoped, lightning struck the kite. It traveled down the wire, electrified the string and caused the hairs of the string to stand on end. Then the electrified twine struck the key and produced sparks. Franklin was thrilled. He proved what he had set out to prove—that lightning is simply one big electric spark.

A curious phenomenon sometimes occurs between the masts of a ship. It appears to be a flash of lightning, but it looks like a blue candle flame. On stormy nights, sailors who saw this eerie light above their ships thought they were witnessing a supernatural appearance, perhaps the dance of water spirits.

St. Elmo's fire is caused by an electrical discharge from a sharp object and is visible only during thunderstorms and only in the dark. Sometimes there is one flame and other times there are two. These flames can occur in places other than between the masts of ships. St. Elmo's fire can occur on other pointed objects, such as steeples, lightning rods, and the wings of airplanes. This strange phenomenon was named after St. Elmo of the Middle Ages who was considered the patron saint of sailors. Some sailors believed that the light indicated St. Elmo's presence—either to assure them of better weather ahead or to warn them of upcoming danger.

PROJECT IDEA

Superstitions arose around St. Elmo's fire just as they arose around lightning that struck at any place and time. Read the following list of superstitions about lightning aloud to the class. Ask students to identify how and why they think each superstition arose and whether or not it has a basis in fact. The Web sites in the Suggested Reading section provide lists of other common misconceptions about lightning.

Lightning never strikes twice in the same place. (False. The Empire State building in New York City is struck by lightning about twenty-three times a year.)

During a lightning storm, seek shelter in a cave. (False: Caves can have iron and copper in the walls, which attract lightning. Myths often say that lightning is made in underground caves.)

It is possible to disperse lightning by ringing bells. (False: People thought that the sound waves created by the bells deflected the lightning.)

Oak trees offer protection from lightning. (False. This belief arose in ancient times. Because thunder is often heard rustling the branches of oak trees, oak trees became the trees of the Thunder god. Trees are one of the most common objects struck by lightning.)

Rubber shoes and rubber tires offer protection from lightning. (False. Rubber offers no protection. The reason you are protected in the car is because the lightning will hit the metal on the outside of the car, not the inside.)

St. Elmo's fire indicated fair weather ahead. (True: This arose because St. Elmo's fire was commonly seen six hours after the center of a storm passed.)

SUGGESTED READING

Below are two of the numerous Web sites that address common misconceptions about lightning and explain why they are wrong.

ThinkQuest. "Common Misconceptions About Lightning." *http://www.library.think quest.org/16132/html/lightninginfo/myths.html.*

Zavisa, John. "How Lightning Works." *http://www.howstuffworks.com/lightning.htm.*

TOPIC 6. The Sound of Thunder

The principle behind the sound of thunder is much the same as the principle behind the sound of popping corn. When we make popcorn, the corn is heated by steam or warm air. The warm air expands and explodes, and we hear the characteristic popping sound.

PROJECT IDEA

Make popcorn in your classroom and explain the process to students as they listen to the sounds the popping kernels of corn make.

SUGGESTED READING

No suggested reading for this one—just enjoy the popcorn!

—— SUGGESTED READING FOR TEACHERS

National Lightning Safety Institute. *http://www.lightningsafety.com/*.

> Contains lots of information about lightning safety and lightning incidents. Has a list
> of resource material.

3 .... Wind

──────── THE MYTHS OF THE WIND

Wind gods appeared in myths and legends around the world because people understood the significance of wind. Not only did they fear the wind because it could turn fierce and destructive, but they also were in awe of the wind because they connected it with breath and life. The wind moved, and as it did it seemed to breathe life into the world. It stirred the plants and animals, yet it remained invisible.

In legends around the world people believed that sorcerers or wizards brewed the winds in caves and then released them. Early people seemed to believe that the winds were confined, either in caves or in some sort of container, such as a leather sack or a gourd. In most legends and myths, someone controlled the winds, and he or she could release them at will. These wind gods could send favorable winds to help sailors steer their vessels, or they could unleash destructive winds and cause hurricanes, tornadoes, and tempests.

In many cultures, people identified winds associated with the cardinal directions. Personifications of the north wind, the south wind, the east wind, and the west wind are common all over the world, but the Greek winds achieved particular fame. The Greeks personified each wind and gave it a personality based on its characteristics. Winds that blew from the north were usually strong and cold, so the north wind became the blustery Boreas. Winds from the west, on the other hand, were usually soft and light, so the west wind was personified as the gentle Zephyrus who ushered in the warm weather.

Greece had lots of wind gods, but Aeolus was the most famous. Read the following myth of the Greek wind god, Aeolus. Then identify beliefs that the ancient Greeks had about the wind and compare them to the science that emerged years later.

"The Guardian of Winds," a Myth from Greece

A long time ago, Aeolus, the son of the sea god Poseidon, won the favor of the gods and came to inherit the Island of Winds. Aeolus was the guardian of all the winds of the world, and he could release them at will from within a cave nestled deep within his island home. Some say that Aeolus was the father of the directional winds—the winds of the north, south, east, and west. Other say he was the father of all the winds and that he had the ability to rouse them to fury or calm them down. Every sailor who sailed his father's waters knew of Aeolus. Often these sailors called on Aeolus to guide and protect them. If Aeolus got angry, he had the power to release harmful winds that steered the sailors off course and caused shipwrecks. If Aeolus remained pleased, he had the power to release gentle trade winds that helped the sailors navigate smoothly through the waters and arrive safely at their destinations.

One day, one of the best-loved sailors of ancient Greece arrived on Aeolus's island. The name of this sailor was none other than Odysseus, who was once the king of Ithica. Odysseus had launched twelve ships ten years earlier and set out with his crew to fight the Trojan War. He fought bravely in the war and won. Now he wanted nothing more than to return to Ithica and to his wife Penelope and their son, Telemachus.

Odysseus and his crew had been sailing the seas for some time, trying to reach Ithica. They struggled against furious storm winds that steered their ship far off course. Not only did Odysseus encounter storm winds but he also met a host of other treacherous conditions. Odysseus and his men came across rocks and whirlpools, and they found themselves shipwrecked on strange islands—frightening places where they met evil characters and dangers of all sorts. When Odysseus came upon the Island of Winds, he had a different experience entirely. The Island of Winds was a peaceful place, and Odysseus was treated royally by Aeolus, the ruler of the island. Odysseus and his crew remained with Aeolus and his family for a month. They relaxed in comfort and enjoyed the hospitality of Poseidon's kind son. Odysseus soon learned that Aeolus had the power to send him safely home, and Aeolus told the travelers that he had every intention of helping them return to Ithica.

After a month's time Aeolus gave Odysseus a goatskin bag that contained all the winds that would impede his voyage—the very same winds that had terrorized the weary crew for so long and led them into strange and dangerous places. Aeolus kept the winds locked tightly away inside a cave. When Aeolus took Odysseus to the cave, Odysseus saw that the cave had twelve holes, each blocked with a rock. When Aeolus removed one rock, the strong north wind would gust out of the cave, and when Aeolus removed another rock the gentle west wind would blow. When Aeolus removed all twelve of the rocks, he released a hurricane—but he had no intention of releasing a hurricane then. Aeolus had taken a liking to Odysseus, and he confined all the treacherous winds in a goatskin bag and tied them with a silver string. Then he handed the bag to Odysseus.

"Keep these winds locked tightly in the bag," Aeolus warned Odysseus, "for they are the most powerful winds of the world. These winds can sweep over the waters with the force of demons. They can stir up waves that render ships powerless. The most skilled of all sailors have met their deaths in the face of these winds. You have seen for yourself what trouble they can cause. Now these winds are in your power. Keep them confined until you arrive safely in Ithica."

Odysseus thanked Aeolus and left the island, feeling hopeful and very fortunate that Aeolus had bestowed on him this powerful gift. Once Odysseus's ship sailed away from the island, Aeolus removed a rock from the cave and released the gentle west wind to guide the ship. By the grace of the gods, Odysseus and his crew were on their way home.

Now, Odysseus had no intention of angering the gods, and he had every intention of heeding Aeolus's advice. He stayed awake for nine days, and, guided by the gentle west wind, he steered his ship. Finally, as if by miracle, he saw Ithica in the distance. For the first time in years, Odysseus was overcome with joy. In just a short time he would be home. But alas, it was not the will of the gods that Odysseus return home—not then anyway. Odysseus had remained awake to steer the ship since the moment they left the wind god, but with Ithica in sight he finally gave in to exhaustion. While Odysseus slept, his crew got restless. Curiosity got the better of them. What can be inside this bag? they wondered. They knew that the bag had been given to Odysseus by Aeolus. They could only guess that it contained valuable treasure—treasure that could only be had by someone whose powers had been entrusted to him by the great god Zeus.

The crew argued and fretted over the bag for some time while Odysseus slept. They discussed among themselves whether or not to open it. Some of the crew got angry. Hadn't they stood by Odysseus during their long, hard journey? they argued. Didn't they deserve to share in the treasure, too? Finally, the crew gave in to their curiousity and to the anger they felt about being denied Aeolus's gift. They opened the bag, and disaster struck instantly. Winds emerged with powerful force. They pummeled the ship. Enormous waves sprang from the waters and thrashed against the ship. It tossed and it turned this way and that and threatened to capsize.

"We have released a tempest!" one of the crew members cried. Odysseus awoke with a start. Horrified, he watched the wind and the waves, and he realized he had no power against the raging storm. It was all the sailors could do to keep the ship above water. In no time at all they found themselves pushed furiously in the opposite direction of Ithica. They crashed up on the shore of the Island of Winds—right back where they started. Aeolus stood nearby, aghast, staring in disbelief at the dumbfounded sailors.

Once Odysseus gathered his wits together, he begged Aeolus to confine the winds once again and help him get home to Ithica. But Aeolus was not so kindhearted this time. The guardian of winds refused to confine the winds, knowing for certain that Odysseus had angered the gods and there was nothing more a wind guardian could do to help him. Aeolus whisked the unfortunate sailors back into the sea as fast and furiously as they had arrived. It would be ten long years before Odysseus returned home to Ithica, and the raging winds swirled furiously over the waters the entire time.

· · · · · · · · · · · · · · · · · · · ·

"The Guardian of Winds" is based on Homer's *Odyssey,* translated by Robert Fitzgerald (New York: Anchor Books, 1963), and on various synopses of this work.

THE SCIENCE OF THE WIND

The ancient Greeks developed numerous theories about the wind and weather. At first, the Greeks believed that each wind was controlled by a different god, and Aeolus was simply the guardian of the winds. Later, Aeolus became the god of all winds. Aristotle was perhaps the most influential scientist of ancient Greece, and he incorporated his philosophy of the wind into a book called *Meteorologica.* This is where the word meteorology comes from. Aristotle believed that all winds came from the Earth and were produced by the Sun. He stated that the winds were hot and dry in the summer when the Sun was strong and cold and damp in the winter when the Sun was weak. Other Greek philosophers advanced other theories about the wind. Pythagoras believed that air—and by extension, wind—was one of four elements that made up the universe, along with fire, earth, and water. Like breath, air and wind were typically considered the source of life. For this reason, the Greek philosophers considered air the most important element. Air gods were immortal, and they often held high positions among the gods. The great god Zeus and his wife, Hera, controlled the air and all forms of atmospheric phenomena, including wind.

Review the following beliefs that surfaced in the Greek myth of Aeolus. Topics for discussion and projects follow.

BELIEF: There are different kinds of winds.

Early theories about wind and weather worked their way into the myths and colored the way the Greeks personified the various winds they recognized in the world. The ancient Greeks recognized many different kinds of wind in the world: some that moved the air, some that moved the Earth, and some that moved the waters. Greece was a windy place, and the people of Greece understood the importance of wind because it affected their lives every day. In this area of the world, winds came and went at different times of year, and they blew in from every direction. Myths from Greece and other lands often connected wind with the four cardinal directions. In these myths, four gods sat in the four corners of the world, the north, the east, the south, and the west, and from these stations they blew winds all over the Earth. In Greek myths, Boreas personified the north wind: Zephyrus, the west wind: Notis, the east wind: and Eurus, the south wind. The Greeks also had gods that personified the semicardinal directions and many other wind gods who materialized in one form or another and affected the Earth in different ways.

Wind is a feature of the atmosphere, and the atmosphere is composed of layers of air. The exosphere is the layer that's farthest from Earth. Then, next farthest is the thermosphere, then the mesosphere, and then the stratosphere. The troposphere is the layer that's closest to the Earth's surface, and that's where all forms of weather take place. Meteorologists consider the term *weather* to encompass six elements: temperature, pressure, moisture, clouds, precipitation, and wind. All six of these elements can be measured, and they combine in different ways to cause variations in the weather.

Though the winds vary by force, speed, and direction, they all fall into three basic categories: local winds, regional winds, and global winds. Local winds are called convection winds, and they occur when warm air rises and cold air replaces it. Sea breezes are local winds that blow in from the sea, usually during the daytime. This is when the Sun warms the land, and as the warm air rises the colder air over the sea moves inland and replaces the warm air. Land breezes are local winds that blow out to sea, usually in the evening. When the Sun sets and the Earth gets cooler, the warmer air over the sea rises, and the cool air from the land moves outward and replaces the rising sea air. In Greece, where mountains line the coastal areas, two other types of local winds blow: mountain winds and valley winds. These winds blow because of the difference in temperature between the mountains and the valley. In the daytime, the air over the mountains becomes much warmer than the air over the valleys, and at night it's the other way around. Like local winds, regional winds and global winds are caused by temperature differences, but these winds travel much farther than local winds.

BELIEF: Winds can be good or evil.

Winds can be warm or cold, light or strong. People who personified the winds generally considered warm winds and light breezes benevolent and cold, icy winds and strong, forceful winds evil. Benevolent winds that blow from the tropics bring warmth to cold areas of the world, and benevolent winds that blow from the poles cool down the hot desert lands. Evil winds cause hurricanes, tornadoes, and violent storms.

The difference between forces of wind is simply a matter of scale. Tornadoes are much bigger and often more violent than whirlwinds, and hurricanes are much bigger and often more violent than tornadoes. It is easy to see how early people personified these forces of wind. The Greeks and other early mythmakers characterized the light breezes as supernatural beings who were gentle and kind and the most violent winds as giants, dragons, and monsters of all sorts. Sometimes these monsters grew as the storms themselves grew. In Greek myths, Otus and Ephialtes, the giant sons of Poseidon, the sea god, grew nine inches every month. They continued to grow until Apollo, a Greek sun god, shot them with his golden arrows. In other words, when the Sun broke through the storm, the winds died down and disappeared.

People attempted to classify the winds for centuries. Winds were classified by direction, force, and location, but everyone seemed to perceive the force of the wind a little differently. In 1805 Francis Beaufort, an admiral from the British navy, developed a standard classification for the winds. This classification survives to this day. The Beaufort scale has been adapted and is currently used by the United States Weather Service. This scale classifies and names the winds by their velocity. *Wind velocity* is a term that generally equates to wind force but that encompasses both the speed and direction of the wind. Beaufort's scale includes twelve winds in order of velocity, and it identifies each of these winds by how it affects objects on Earth.

BELIEF: Wind is a source of energy.

The ancient Greeks considered wind a source of energy. It moved the world, and we have heard how it moved Odysseus's ship. Wind is a tremendous source of energy, and it comes from the natural movement of the air. Sailors have always relied on the wind to move their vessels, and thousands of years ago people relied on the wind to power the first industries. Windmills came into use to grind grain and pump water and are still used in some places today. In today's world, however, wind turbines have come into common use. Turbines are successors to windmills, and they have blades that rotate in the wind and produce electricity.

Wind power is considered a form of solar energy because wind movement depends on the Sun to send heat to different parts of the Earth. As the Sun heats the Earth, hot air rises and cool air replaces it, and wind distributes energy from warmer areas of the Earth to cooler areas. Wind energy and solar energy are renewable resources, which means that they never run out. Nonrenewable resources, such as coal, oil, and natural gas, will run out eventually, and when they dwindle it will be far too costly to obtain them. Currently, the United States gets most of its energy from nonrenewable sources called fossil fuels. Fossil fuels include coal, oil, and natural gas that exist underneath the Earth's surface. Over time, extracting these fossil fuels hurts the environment. For this reason environmentalists and other people concerned with protecting the Earth push for scientists to investigate ways to extract most of our energy from the Sun and the wind.

BELIEF: Winds can be unpredictable.

In Greece the summers are hot and dry, and in the winter it rains. The land is surrounded on three sides by bodies of water: the Mediterranean Sea, the Ionian Sea, and the Aegean Sea. The Aegean Sea has numerous arms that cut into the land and cause unpredictable winds that roll in from every direction, particularly in winter. For this reason, the ancient Greeks came to believe that the winds had minds of their own. For a long time, none of the numerous Greek wind gods had total control. Scholars have theorized that this probably reflects the country's erratic wind patterns. In "The Guardian of Winds," and in the *Odyssey* from which it is derived, Odysseus and his crew had no control over the winds until Aeolus helped them. Odysseus's ship was tossed to all areas of the sea until the crew finally met the man who could help them. Even Aeolus knew that Zeus and the other wind gods had a say in which winds blew, however. When Odysseus and his crew let the winds out of the bag and were blown back to Aeolus's island, the Guardian of Winds told the crew that he could no longer help them because they had obviously angered the gods.

BELIEF: Winds can help sailors.

In "The Guardian of Winds," Aeolus sent the trade winds to help guide Odysseus back to Ithica. Sailors had many myths and legends about the wind because they depended on the wind to sail their vessels. In fact, in some areas of the world, disreputable people capitalized on this dependence and built businesses selling trade winds to sailors. Trade winds are low level winds that blow consistently in one direction. In the Northern Hemisphere, they blow from the northeast, and in the Southern Hemisphere they blow from the southeast. Trade winds are caused by a series of rotating air cells that move back and forth from the equator to the poles. In the Northern Hemisphere, the air moves from the north toward the equator, but because of the Earth's rotation the air curves. This makes the wind blow from the northeast, and it blows in a soft, steady flow.

BELIEF: Wind can be harmful to sailors and hinder their voyages.

In some regions on Earth, the winds blow predominantly in one direction throughout the year, and in other regions the winds change with the seasons. Of course, the winds can vary from day to day depending on other factors, such as temperature and rainstorms. This makes it hard to detect the prevailing wind direction in most places. Sailors learned the way of the winds because they had to learn it. In ancient times they traveled dangerous waters with nothing but sails to move them through the water. This left the sailors totally at the mercy of the gods of the sea and the winds. In the waters around Greece, sailors had to learn which winds prevailed during different seasons. In Greece cyclones or hurricanes follow the prevailing winds. These cyclones can crop up unexpectedly and terrorize anyone attempting to navigate the seas.

In temperate regions of the world, the sharp contrast in temperature between the water and the land affects the air pressure. This contrast in temperature is what causes the monsoon winds. In these areas, winds tend to blow heavily over the land in the winter because the land is so much colder than the water. In summer, when the water is colder than the land, the winds tend to blow most heavily over the seas. In the case of monsoons, the cold water cools the air above it, and, when the cool air blows over the seas, it eventually becomes too heavy to hold moisture. This results in heavy rains that are swept toward the coast by powerful winds.

— TOPICS FOR DISCUSSION AND PROJECTS

TOPIC 1. The Wind in Greek Philosophy

The Greek philosophers had much to say about the wind and weather. We already mentioned a few of the early theories, but there are many more. The theories and discoveries the ancient Greeks had about the natural world had a tremendous influence on the history of science.

PROJECT IDEA

Have students write a biography on one of the Greek philosophers. Tell students to make the scientific theories of the philosopher the focus of their biographies and ask them to include in their biography ideas the philosopher had about wind and weather. The following books provide information about Greek philosophers and their scientific theories. You might want to direct students to the reference section of your school library for additional biographical information. The librarian can help students locate information in the *World Book Encyclopedia* and specialized biographical dictionaries and encyclopedias.

SUGGESTED READING

Anderson, Margaret. *Scientists of the Ancient World*. Springfield, NJ: Enslow Publishers, 1999.

> Includes the theories of ten prominent scholars of ancient times, including, among others, Pythagoras, Pliny, Archimedes, and Eratosthenes.

Gay, Kathlyn. *Science in Ancient Greece*. New York: Franklin Watts, 1998.

> Discusses the scientific theories and discoveries of Ptolemy, Pythagoras, Hippocrates, and Aristotle.

Nardo, Don. *Scientists of Ancient Greece*. San Diego, CA: Lucent Books, 1999.

> Includes information on the science of Democritus, Plato, Aristotle, Theophrastus, Archimedes, Ptolemy, and Galen.

Parker, Steve. *Aristotle and Scientific Thought*. New York: Chelsea House, 1995.

> Concentrates on the scientific thought of Aristotle, including his theories on wind and weather.

TOPIC 2. Weather Vanes

Weather vanes symbolize our dependence on the wind and weather. They were invented long ago because people needed to know the direction the wind blew. That's how the winds got their names. Weather vanes always point toward the direction from which the wind blows.

The oldest weather vane that scholars have discovered sits above the Tower of Winds in Athens, Greece. Triton, the Greek Sea god, takes the place of the familiar rooster that adorns many weather vanes today. Triton is half man and half fish, and in myths he holds a trident, which is a three-pronged wand, in his hand. The bronze statue of Triton that sits on the Tower of Winds is large, and Triton points in the direction of the wind. In Greek myths, sea gods often had the ability to rouse the winds that blew over the waters.

PROJECT IDEA

Have students make a weathervane using a wire coat hanger. Follow the instructions below to help them complete their projects. You might want to copy the instructions on the board or pass out a copy to each student. This project is not difficult, but it can take some time to complete. Allow about an hour to complete it if you are doing the project in class.

How to Make a Weather Vane

Materials

A wire coat hanger (must be all wire)
Paper
Tape
Paper clips
String

Steps to Follow

1. Have students place their coat hangers on a piece of paper and draw a straight line from the hanger's neck to the bottom of the hanger. Then have them trace around one corner of the hanger to make a triangle and cut out the triangle.

FIGURE 3.1 · A Weather Vane (The heavy side of the weather vane will move to the direction from which the wind blows.)

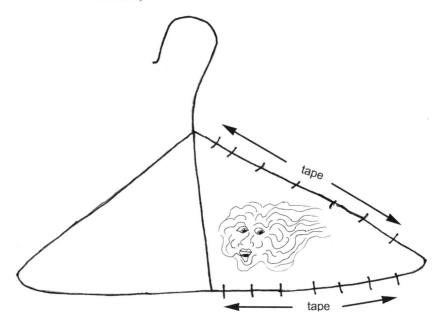

2. Have students decorate their papers on both sides. They can color a picture of a rooster on their paper, or they can choose another design. Tell them to choose a design they consider a symbol for wind and initial their designs.

3. Tell students to tape the paper inside the hanger where they cut it out. Make sure they wrap the tape around the outside wire to keep the paper secure and make sure there are no gaps.

4. Using the poster board, have students cut out the letters *N, E, S,* and *W* using the stencils you provide.

5. Students can then hang their weather vanes in a tree. If there is a tree outside your school, you might to want to hang the weather vanes there. If not, students can hang their weather vanes in trees at home. Tell students to make sure they hang their weather vanes on a sturdy branch that is not swayed by the wind and to tie the vanes in place with string. The weather vanes should be completely level. If they are not, students can tape paper clips to the high end of the hanger to balance it.

SUGGESTED READING

Kinderart. "Folk Art Weathervane." *http://www.kinderart.com/folkart/weathervane.shtml.*

> Has instructions for creating a weather vane and provides information on the history and use of weather vanes.

Lunde, Anders. *Action Whirligigs*. Mineola, NY: Dover, 2003.

> One of a series of illustrated books by Lunde that contains patterns and instructions for making inexpensive and easy wind-powered toys. Other books in the series include *Easy to Make Whirligigs* (Dover, 1996), *Making Animated Whirligigs* (Dover, 1998), and *Whimisical Whirligigs* (Dover, 2000).

Woods, Bruce, and David Schoonmaker. *Whirligigs and Weathervanes: A Celebration of Wind Gadgets with Dozens of Creative Projects to Make*. New York: Sterling, 1992.

> Contains instructions for numerous creative projects suitable for students, including weather vanes, whirling ducks and airplanes, and numerous wind-powered toys. Has color illustrations.

The Beaufort Scale of wind force categorizes winds from light breezes to hurricanes. Each of the twelve winds have distinct characteristics, which led the Greeks and other people in the ancient world to give them distinct personalities.

PROJECT IDEA

Review the winds on the Beaufort scale in class or make a copy of the scale to distribute to students. Have students choose one of the winds on the scale and create a character to personify the wind. Tell students to think about the characteristics and effects of the wind they wish to personify. This will help them create an appropriate character. Have them write a short character description of their wind personality. The following list of Greek wind gods might help students think of ideas for their own characters.

Wind Gods of Ancient Greece

Athena—Athena had many functions in Greek mythology, and in one aspect she was the goddess of the air. She could create or calm the storm winds, and she often controlled the air or wind that helped sailors sail their vessels.

Aura—Aura was a winged air nymph who controlled the morning wind and the gentle breezes.

Boreas (Aquilo in Roman myths)—Boreas was one of the winds of the four cardinal directions. He personified the strong and blustery north wind. He also gathered the clouds and scattered them over the sky.

Eurus—Euros was the god of the south wind. His name is the same in Greek and Roman myths.

The Harpies—The Harpies were winged genii who represented the storm winds. There were three Harpies: Aello, Ocypete, and Celaeno. The term *Harpy* derived from the Greek work *harpazein,* which means "to snatch or carry away." In myths, the Harpies were fierce and evil beings who had the heads of humans and the bodies of birds. They swooped from the sky and snatched food and human beings off the ground, and they spread filth and debris across the land.

Hera—Hera was the wife of the sky god Zeus and the goddess of the atmosphere. Like Zeus, she had the power to control the storm, thunder, and wind.

TABLE 3.1 · The Beaufort Scale

Type of Wind	Observed Effect	Miles per Hour
Calm	Smoke rises vertically.	Less than 1
Light air	Smoke shows wind direction.	1–3
Light breeze	Wind is felt on face. Leaves rustle. Weather vanes show wind direction.	4–7
Gentle breeze	Leaves move constantly.	8–12
Moderate breeze	Wind moves dust, paper, and small branches.	13–18
Fresh breeze	Small trees sway. Small waves form on water.	19–24
Strong breeze	Large branches move.	25–31
High wind	Whole trees move.	32–38
Gale	Twigs break off trees.	39–46
Strong gale	Large branches break.	47–54
Whole gale	Trees are uprooted.	55–63
Storm	Widespread damage occurs.	64–73
Hurricane	Extreme damage occurs.	74 or more

Hermes—Hermes was a son of Zeus and the winged messenger who flew with great speed to deliver messages from the gods. In myth, Hermes often controlled the swift and strong north wind.

Notis (Auster in Roman Myths)—Notis was one of the winds of the four cardinal directions. He personified the south wind that frequently brought rain.

Otus and **Ephialtes**—Otus and Ephialtes were the giant sons of Poseidon, the sea god. They personified the hurricanes, and like the hurricanes they started out small and grew to enormous size. Some myths say that Otus and Ephialtes grew nine inches every month.

Pan—Pan was the Greek god of pastures who protected the sheep and goats. He had the body of a man and the horns and legs of a goat. Pan was a gentle and easygoing god who stirred the gentle breeze when he blew on his flute.

Poseidon—Poseidon was the ruler of the sea and one of the three great gods of ancient Greece, along with Zeus, who ruled the sky, and Hades, who ruled the land underneath the Earth. Poseidon had a vicious temper and often raised deadly hurricane winds from the ocean depths.

Typhon—Typhon was a dragon that personified the hurricane or the typhoon. He lived in the Corycian cave in Sicily, and he had hands made of a hundred serpent heads, eyes that flashed lightning, and wings so large they covered the sky in darkness. Typhon was the largest monster ever born; he was taller than the trees and he spat fire and vomited rocks. He terrorized all the gods of Olympus and often caused them to change shapes.

Zephyrus (Favonius in Roman myths)—Zephyrus personified the gentle west wind. In myth, Zephyrus often ushered in the spring and fair weather.

TOPIC 4. Air Pressure and Weather Prediction

Ask students if they have ever been on an airplane or on a high mountain. If they have, they may know what it feels like when their ears pop. Tell them that this is because as they move higher the temperature decreases. The air pressure also decreases, which makes their ears pop. Ears pop in an attempt to balance the air pressure inside and outside the ears.

Weather forecasters know the relationship between temperature and air pressure. They use barometers, which measure air pressure, to predict the weather. High air pressure indicates the coming of cooler temperatures and clear skies. Low air pressure indicates the coming of warmer weather and storms. Before hurricanes and tornadoes, air pressure drops considerably. This is when weather forecasters today issue storm warnings. Long ago, people who had no barometers had to rely on their keen sense of observation to predict the weather.

PROJECT IDEA

Have students make a barometer they can use to forecast the weather using the following directions.

Materials

Clear plastic drinking straws
Glass bottles
Something to plug the bottle, such as putty or dough

How to Make a Barometer

1. Insert the drinking straw into the bottle. (The straw should extend a ways out of the bottle, and the water level in the straw should extend a ways out of the bottle.)

2. Fill the bottle halfway full of water.

3. Add a few drops of food coloring to color the water in your barometer.

4. Seal the neck of the bottle around the straw with putty.

Explain to students what happens. If the air pressure outside the bottle decreases, the water will rise in the straw. If the air pressure outside the bottle increases, the water will drop in the straw.

SUGGESTED READING

Friend, Sandra. *Earth's Wild Winds*. Brookfield, CT: Twenty-First Century Books, 2002.

Examines various aspects of the wind, including its effects and measurement.

NASA. "It's a Breeze: How Air Pressure Affects You." *http://kids.Earth.nasa.gov/archive/air_pressure/.*

Discusses air pressure, how it affects us, and how it is used to predict the weather. Has instructions for making a barometer.

Rodgers, Alan. *Wind and Air Pressure*. Chicago: Heinemann Library, 2003.

Discusses air pressure and barometers as well as other technological developments for measuring wind.

TOPIC 5. The Tower of Winds

The Tower of Winds is a monument that was constructed in Athens, Greece, as a temple to Aeolus. The Tower of Winds is also called the Horologion, and travelers to Athens today can see the Horologion in the Acropolis. The Tower of Winds is an octagonal structure made of marble, and each of the eight sides faces one of the principal directions. In ancient times, the Greeks used the Tower of Winds to keep time. The tower had a sundial and a water clock to measure time and a weather vane to monitor wind direction. Each of the eight sides of the Horologion has a figure that represents one of the directional winds.

PROJECT IDEA

Have students create a replica of the Tower of Winds out of clay or papier-mâché. Tell them to use symbols or pictures to represent the directional winds. To create their towers, students should use what they learned about the characteristics of wind and wind gods connected with each direction. The books in the Suggested Reading section are just a few of the many sources that contain instructions for making models from papier-mâché and clay.

SUGGESTED READING

Haldane, Suzanne. *Faces on Places: About Gargoyles and Other Stone Creatures*. New York: Viking Press, 1980.

> Discusses the work of artists who carved faces and figures on ancient buildings.

Schwartz, Renee F. *Papier-Mâché*. Toronto: Kids Can Press, 2000.

> Includes eleven projects and lots of tips for making successful papier-mâché projects.

Seix, Victoria. *Creating with Papier-Mâché*. Farmington Hills, MI: Blackbirch Press, 2000.

> Contains fifteen projects and instructions for creating with paper strips and paper pulp.

Weiss, Harvey. *Model Buildings and How to Make Them*. New York: Crowell, 1979.

> Contains instructions for creating model buildings out of cardboard and wood.

TOPIC 6. The Winds in Greek Literature

In the Greek myths, supernatural beings who controlled the wind lived in the Earth, the sea, and the sky. The Greeks witnessed the winds in all three realms; they felt the wind in the air, they saw how the winds affected objects on Earth, and they knew how the winds affected ships at sea. In the Greek epic *The Odyssey*, Odysseus had encounters with many supernatural beings as he sailed the seas, and many of those beings personified elements of weather. Many scholars believe that all the Greek and Roman gods controlled the elements and that they originally personified the forces of nature. The highest gods personified the strongest forces, and the lesser gods controlled the weaker forces.

PROJECT IDEA

Have students read a children's version of Homer's *Odyssey* or read chapters of *The Odyssey* aloud in class. The following list of books represent only a few of the numerous versions of *The Odyssey* available for young readers. Your school or library may have others. Have students choose one of these chapters in *The Odyssey* and write a summary of that chapter. Tell students to look for symbols of the wind and other meteorological phenomena in their chapter and to include those symbols in their summaries.

SUGGESTED READING

Mattern, Joanne. *The Odyssey*. New York: Harper, 1996.

Mitchell, Adrian, and Stuart Robertson. *The Odyssey*. New York: Dorling-Kindersley, 2000.

Sutcliff, Rosemary, and Alan Lee. *The Wanderings of Odysseus: The Story of the Odyssey*. New York: Delacorte Press, 1966.

—— SUGGESTED READING FOR TEACHERS

Hopping, Lorraine Jean. *Today's Weather Is . . . A Book of Experiments*. New York: Mondo Publishing, 2000.

> Contains news stories about the weather and instructions for related experiments students can make with common household items. One of the stories focuses on a school that runs solely on wind power.

Kwitter, Karen. *Atmosphere and Weather*. Portland, ME: J. Weston Walch, 1998.

> Contains activities on the science of the weather and the weather systems of the world.

Maton, Anthea. *Exploring Earth's Weather*. Englewood Cliffs, NJ: Prentice Hall, 1994.

> Explores meteorology and climatology. Includes an annotated teachers' edition, activity book, lab manual, and test book.

University Corporation for Atmospheric Research. *http://www.ucar.edu/educ_outreach/k-12.html*.

> Contains links to numerous sites of teacher materials on atmospheric science.

4 ·········· Fogs, Mists, and Mirages

THE MYTHS OF FOGS, MISTS, AND MIRAGES

Fog and mist can be eerie. They appear to drift and change shape, and as they do images materialize within them and conjure up visions of mysterious lands. The shapes that appear out of floating fogs and mists gave rise to stories about mystical beings and elusive worlds hidden from view. Dense fogs can fool the eye and create optical illusions. Mountain paradises, undersea palaces, and phantom ships have all appeared to materialize within fogs. This led to legends of enchanted castles and mystical realms inhabited by immortal beings of all sorts, including fairies.

In British legend, just such a mystical place exists, and it is the abode of Morgan le Fay, the fairy queen of King Arthur's Camelot. In some legends, Morgan's castle sits on a mist-covered island in the Atlantic Ocean at the end of the silver path lit by the Moon. In other legends, the castle is under water in the Strait of Messina, somewhere between Italy and Sicily. People have seen castles floating above the water in the Strait of Messina, and they believed that the fairy queen created these castles by magic. The castles appear to rise out of the sea and change shape. Science has proved them to be an illusion, but the legend of a mystical land survived for many years.

Read the following legend of Morgan le Fay, the fairy queen of the Middle Ages. Then compare the legend to facts about fogs and mists and a related phenomenon called the mirage. Topics for discussion and projects deal with both the atmospheric conditions that produce these phenomena and with legends from the British Isles.

"THE ENCHANTED CASTLES OF MORGAN LE FAY," A LEGEND FROM BRITAIN

Long ago in the Middle Ages there lived a beautiful fairy named Morgan le Fay. She lived in the days when King Arthur ruled Britain. Morgan was Arthur's half-sister, and she was as powerful as she was beautiful. Morgan le Fay learned the art of magic from Merlin, a famous magician in King Arthur's court. Merlin served as Arthur's advisor, and he helped Arthur build a kingdom that he ruled for many years from his castle of Camelot.

Morgan le Fay was a great healer and a protector of Arthur. She visited the castle of Camelot often, but as fairy queen she had castles of her own. Morgan had castles in the British Isles and on the island of Avalon where she lived with her eight fairy sisters, but she also had a castle under the sea and numerous castles that floated in the air. Morgan's Avalon was surrounded by mist, and her castles under the sea were made of shimmering crystal.

Now, the castles of a fairy queen were bound to be enchanting, but the castles that floated above the sea were undecidedly magic. Anyone who saw these castles had a hard time believing that they were suspended in air. People found it curious that these castles appeared to materialize out of nowhere and could never be reached. For a long time people did not know whether these castles truly existed or if they existed only in the fairy realm—the realm that so often appears in legends and that weaves its thread through the real world and materializes in one form or another at unexpected times.

Legends say that Morgan le Fay lived in this magical realm. She was a fairy after all, and she had all the abilities typically attributed to fairies, including the ability to fly and to change shape. Legends say that Morgan le Fay was a poet and a musician and that she had the power to heal the battle wounds of knights who fought in King Arthur's court and who came close to death. Morgan had these abilities, legends say, because she understood the ways of the Earth. She knew the healing properties of all the herbs and plants that grew abundantly in Avalon. Then under Merlin's tutelage she learned the art of sorcery. When Arthur was near death, he was taken to Avalon. There, Morgan miraculously healed the famous king—whether or not it was solely by magic that she healed him no one knows for sure.

There are numerous legends of the fairy Morgan that became entrenched in the literature of Britain long before Arthur took rule. Fairies are timeless and ageless, and stories of magical women and mist-covered castles have been told in the British Isles for as long as anyone can remember. In some of the earliest stories, a magic woman much like Morgan lived under the sea and kept the company of mermaids. Many people feared Morgan and her supernatural companions. Mermaids were beautiful but extremely dangerous. They lived in underwater palaces made of gold and crystal, and they lured anyone who came near the water down under the sea, where they remained captive forever.

Legends of mermaids and fairy queens were passed down from generation to generation and from place to place. As the legends traveled, they changed, and by the time they arrived in Italy Morgan le Fay may or may not have had anything to do with the evil mermaids who drowned human beings. In any case, however, Morgan le Fay had castles, and from time to time people on the coast of Italy saw her castles drifting silently over the water. Those who heard the legends steered clear of the area, but those who did not hear the legends tried to reach the castles. They never came close. If any human being neared the castles, the castles vanished—as quickly and mysteriously as they had appeared.

It was odd that not just one but many people saw the castles floating mysteriously over the sea. Scholars and learned men of science saw them and even they attributed them to magic. No one could explain it, but no one could deny that in the Strait of Messina, a body of water that separates Sicily from Italy, an entire city of castles seemed to exist but never touched ground. This city looked so strange and mysterious that people knew it must be a product of Morgan's magic. People who heard her legends and knew of her ability to change shape came to believe that this clever fairy could change the shape of other objects in a similar way. It was well known at the time that fairies liked to play tricks. When people saw the floating castles, they said that Morgan conjured up these castles from her abode under the sea strictly for entertainment.

The Strait of Messina was a treacherous place in ancient times. Strong, swift currents kicked up the water unexpectedly, and violent winds frequently threatened to capsize ships. Sailors attempting to navigate the Strait of Messina frequently found themselves battling the elements. So the Strait of Messina was entrenched in superstition as it was. Sometimes, however, the strait seemed miraculously calm. That's when the castles appeared, and that's when people were reminded that fairies truly exist.

Not everyone can see the floating castles of Morgan le Fay; only people who happened to be looking toward Sicily from the coast of Calabria could see them. Even there people only saw them from time to time, and only on warm days with little wind. These strange apparitions extended deep under the sea and towered high into the air. They appeared to be distorted, like castles from another world. What could possibly be responsible for such things? the people wondered. No castle they ever saw in the real world floated in midair. People who knew the ways of mermaids and fairies could only explain it as enchantment. Morgan le Fay was the fairy queen, and she could convert ordinary cliffs and cottages into magnificent palaces simply by force of will. It made sense that no one could ever reach these places—no one can ever penetrate the realm of a fairy, not unless a fairy herself chooses to take them there. But Morgan le Fay, we must remember, is the fairy queen and not just an ordinary fairy or mermaid, and she learned her magic from the most powerful magician in King Arthur's court. Though it would certainly be within her power to take someone to her castle and to allow them to float above the water in the mystical world of the fairy, she does not admit mortals to her world readily. She is as mysterious and elusive as the mists of Avalon. She has powers that can never be discovered by human beings, and she holds all the secrets of the natural world in some misty place in our dreams. It is there that she can conjure up castles that float silently above the waters, and it is there that she can make them vanish into thin air.

. . . .

"The Enchanted Castles of Morgan le Fay" was created from various accounts of the mirage in the Strait of Messina.

THE SCIENCE OF FOGS, MISTS, AND MIRAGES

Many people over the years have seen the enchanted castles in the Strait of Messina. Legends say they belong to Morgan le Fay, but science tells us they are simply a mirage. A mirage occurs under atmospheric conditions similar to those that produce fogs and mists. A mirage is an atmospheric phenomenon that occurs when light passes through layers of atmosphere that differ in temperature and density. The particular kind of mirage that occurs in the strait is called the Fata Morgana, which means Fairy Morgan. This type of mirage usually occurs over cold water and ice because the cold water or ice cools the air directly above it. Sunlight bends more when it travels through cool, moist air than it does when it travels through warm, thin air, but in our mind's eye the light doesn't bend at all. In the Fata Morgana the image that appears is actually there in some form, but it's distorted. The image appears higher and looks much different than it does in real life. It is true that for a long time even scientists thought these images appeared by magic. Like fairies, nature can play tricks on people, and it is actually possible to see images that are not really there.

BELIEF: Fog and mist float.

Fog and mist seem magical because they hover in the air. People who had no idea what caused these phenomena readily perceived them as enchantment, obviously the work of fairies or some other sort of supernatural being who had the ability to conjure up vapors by magical means. Fog and mist are composed of water droplets that drift in the air like clouds. Like clouds, fog and mist appear to float because they are not heavy enough to fall. They hang in the air and often lie close to the ground or the sea or in mountain valleys.

Mist and fog form in the same way clouds form, simply lower to the ground. When warm air moves over cold ground or cold water, the air above the ground or the water becomes cooled, and the water vapor in the air turns to droplets. Fog occurs most commonly early in the morning or late at night. This is because during the night when the sun goes down the ground cools and so does the air just above the ground. Both fog and dew form when water in the air condenses. The temperature at which water vapor begins to condense is called the dew point. Dew forms when water condenses on the ground, and fog forms when water condenses in the air. Mist is simply a thin layer of fog. Fog and mist typically disappear in the morning as the Sun heats up the air, but in the meantime they get trapped in mountain valleys, above marshes, and in low lying areas, and they appear to float.

In the story of the enchanted castles, Morgan le Fay had a castle in Avalon that was surrounded by mist. The mists that float around Avalon are actually sea fogs. Sea fog forms when warm air moves over cold ocean water, and they occur frequently in the waters near the Arctic lands and in cold areas of the Atlantic Ocean. Sea fog is just one of the types of fog that forms in Earth's atmosphere. Take a look at Table 4.1 to see how the four basic types of fog form. You might want to draw this table on the board and fill it in as you discuss with students what characterizes the different types of fogs.

BELIEF: Fog and mist surround islands of the sea.

Fog can form whenever there is moist air, clear skies, and light winds. It can hover in valleys and over lakes and ponds or marshes and it can hover over cold seawater. Fog is formed when warm, moist air becomes

TABLE 4.1 · Types of Fog

Type	Cause
Radiation fog (ground fog)	Cooling close to Earth's surface drops temperature to below dewpoint.
Avection fog	Warm, moist air from the south moves over colder land masses.
Evaporation fog (sea fog)	Cold air moves over warm bodies of water.
Upslope fog (mountain fog)	Wind flows upward and cools air around the mountains as it rises.

so cool that it can no longer hold moisture in the form of water vapor. The air squeezes out the water and forms droplets. These droplets ride on air currents and they drift—as silently and mysteriously as if they were brewed by magical means. Fogs that contain only a few droplets of water appear as thin layers of mist. Fogs that have many droplets of water can be so thick that they have been likened to pea soup.

The mist-covered Avalon is not unlike other depictions of fairy realms or island paradises. These legendary places are often covered with mist because mist is usually symbolic of mystery. Avalon is similarly called Fortunate Isle or the Island of Apples and often the Island of the Blessed because it produces everything its inhabitants need. On this mist-covered island, nature takes care of itself. The fields miraculously cultivate themselves with no help from human hands. The land is eternally fertile, and the crops are eternally abundant. Legends say that Morgan rules this land and that she has knowledge of all the herbs that grow there and how they are used for healing. Fed by the mystical associations of mist and fog, the legendary land of Avalon is pure enchantment. There are scholars who believe, however, that Avalon was a real place. This place, called Glastonbury, is a fertile area of Britain that was once covered with marshes and surrounded by mists. According to some legends, Glastonbury was once a magic island surrounded by water and, like Avalon, a classic example of an island paradise. Avalon is a place where immortals live, a place where no pain or sickness exists, a place where everyone is eternally happy and content.

Fog in itself is not strange and mysterious, but the atmospheric conditions that produce fog can produce a strange and mysterious image. This image is called a mirage, and a mirage is nothing but an optical illusion. A mirage is an inverted image caused by the reflection and refraction of light. The image is not real, but it looks real. Travelers on land often see mirages on long stretches of highway or on desert sand, and travelers on the sea often see mirages over the water. Mirages over the land look like pools of glistening water. Mirages over the sea look like castles or mountains or entire cities floating in air or like phantom ships sailing in the distance.

False images that occur on warm surfaces, such as highways or desert sand, are the most common type of mirage. People who see this kind of image think they see sparkling pools of water when all they really see are rays of light that appear blue because they have been refracted upward by a layer of hot air. This type of mirage is called an inferior mirage. Inferior mirages occur when a layer of cold air moves over a layer of warm air. Cold air over a warm surface reflects light from the clouds. The sky acts like a mirror, and the currents of air moving in the sunlight create an illusion of shimmering water. Mirages that appear in the desert and on highways occur most commonly at sunrise and sunset. A superior mirage occurs above the water, and this type of image is what looked like the crystal castles of Morgan le Fay in the myth we just read. Superior mirages form when a layer of warm air moves over a layer of cold air. Both inferior and superior mirages occur when there are two layers of air lying next to each other and there is a sharp contrast in air temperature between them. The contrast in temperature means that the different layers of air have different densities.

BELIEF: Fog shrouds mystical castles and mysterious
realms.

The legend of Morgan le Fay arose from a type of superior mirage called the Fata Morgana. The Fata Morgana occurs frequently in the Arctic lands and in the Strait of Messina, a body of water between Italy and Sicily. The Fata Morgana is simply a type of mirage that appears under certain atmospheric conditions. Mirages trick the eye, and in legend fairies were always playing tricks on people. The Fata Morgana looks like a floating and vanishing island. Islands have always held fascination for sailors. They hid in the fog, they glistened in the distance, and they seemed to materialize out of nowhere. Mystical islands appeared and vanished in the minds of sailors because, after long months at sea, these seafarers so longed for land that they thought land to be unattainable. That's how islands came to symbolize paradise. Many islands appeared to exist, but they didn't. They were simply a mirage.

The Fata Morgana forms when layers of hot and cold air move over cold water. Light bends toward the surface of the water, and rippling waves and distant trees become distorted by the layers of air and give the appearance of something strange and mysterious hovering over the horizon. The bending light makes the rippling waves or the trees appear higher than they actually are. Then, when the wind blows, it vibrates the images and distorts them. Often the images appear both higher and taller than they actually are. In the case of the Fata Morgana, the distortion forms what appears to be turrets and towers. In some areas of the world, weary sailors saw these images and longed to believe that an island paradise existed somewhere in front of them. In the Strait of Messina, people swore they saw castles, and the castles looked as if they were floating in air.

The Fata Morgana is one type of mirage that occurs from atmospheric conditions similar to those that produce fog and mist. For the Fata Morgana or any type of superior mirage to occur, the air at the surface must be colder than the air above it. A superior mirage forms over something that actually exists, such as trees or buildings or certain land features visible on the horizon. One of the most common occurrences of a superior mirage is the Sun lying on the horizon. The Sun is not really lying on the horizon; it has sunk below it. The reason it looks like it is lying on the horizon is because the sunlight was refracted or bent by the atmosphere. How much the light bends depends on the density of the air.

The term *Fata Morgana* has become a label for a certain type of superior mirage that occurs in several areas of the world, but it originally referred to the mirage seen in the Strait of Messina. In this case the image forms from the distortion of cliffs and buildings along the Sicilian banks. The mirage can only be seen from the tip of Italy, in Calabria, and only on days when there is warmth, sunlight, and little wind. There are other types of mirages similar to the Fata Morgana, and these too have been attributed to supernatural beings. One type of mirage is called the fata bromosa, or the fairy fog. This type of mirage appears fuzzier than that of the Fata Morgana, and it often looks simply like a fog bank hovering over the sea.

BELIEF: You can see strange images in the fog and mist.

We have already discussed the phenomenon of the mirage and the atmospheric conditions that produce it. Floating castles and phantom ships can be frightening to people who don't understand what causes them, but imagine how frightening it would be to see the shadow of a person in the fog standing right in front of you! A phenomenon called the Brocken specter makes this entirely possible. Sometimes, when the fog is in front of you and a light is behind you, your own shadow can be projected in the fog and appear gigantic. Often, the shadow is surrounded by rings of colored light that look similar to a rainbow. Such a specter can look absolutely terrifying. The Brocken specter gets its name from a mountain peak in Germany where the phenomenon has often been seen, and predictably such an occurrence gave rise to legends of demonic monsters. In truth, the Brocken specter is just another of nature's illusions.

Certainly the interplay of light and fog can produce strange and mysterious images, but in dense fogs we see nothing at all. This makes fog one of the most dangerous atmospheric conditions. It is silent and enchanting, but it poses a grave danger to travelers. Fog impairs the vision of travelers along air, land, and sea. Sometimes people speak of a fog as thick as pea soup. Dense fogs have caused accidents on highways and dense sea fogs have caused shipwrecks.

BELIEF: Mist and fog have an evil side.

Many legends portray Morgan le Fay as a good fairy who helps Arthur during his entire reign as king of Britain. Other legends portray her as an evil sorceress. Typically, in legends fairies had both a good and an evil side. The terms *Fata Morgana* and *fata bromosa* draw strong connections between fairies and atmospheric phenomena. In legend, fairies have often been responsible for treacherous conditions, such as fogs, mists, and strong winds.

Dense fog can certainly be treacherous, but a particularly treacherous form of fog is smog. Smog is a term given to foggy smoke that has become a serious form of pollution in large cities. Smog looks somewhat like fog, but it contains particles of smoke and dust. Smog forms when cars and factories release large amounts of chemicals into the air. When those chemicals combine with sunlight, solid particles form and then hang in the air. Mexico City and Los Angeles are two cities that have serious problems with smog, but air pollution from cars and factories has become a global problem. Smog contributes to what has commonly been called the greenhouse effect. When gases get trapped in the Earth's atmosphere, it keeps the Earth warmer than it would normally be. Environmentalists attempt to reduce the level of smog and air pollution and thereby reduce the greenhouse effect. Environmentalists are becoming increasingly concerned with global warming. Due to the greenhouse effect, many scientists believe that the Earth is getting increasingly warmer.

— TOPICS FOR DISCUSSION AND PROJECTS

TOPIC 1. Fairy Fogs and Mirages

In stories throughout the world, fogs and mists typically shroud elusive lands and magical islands. Sometimes these lands were real, and sometimes they were imagined. Since sailors first began sailing the seas, they told myths and legends about islands. Many of the legends attempted to explain what made these lands seem like paradise and who might live in these mystical places.

PROJECT IDEA

The following Web site contains pictures of the Fata Morgana, the fata bromosa, and other mirages. Have students paint a picture of a magical land that might appear to be hidden within a fog or a mist. This is a good art project for students studying light and shadow. Viewing the pictures on the Web site will help students get ideas for their art. The books listed below should help students learn how to use light and shadow in their paintings.

SUGGESTED READING

Court, Rob. *Light*. Chanhassen, MN: Child's World, 2003.

> Gives examples of light and shadow used in art throughout history.

Dimdima Kids. "The Strange World of Mirages." *http://www.dimdima.com/science/Quiz/show_quiz.asp?q_aid=8*.

> A science site that lists and explains different mirages around the world.

Richardson, Joy. *Using Shadows in Art*. Milwaukee, WI: Gareth Stevens, 2000.

> Explains how artists use light and shadow in their paintings and helps students learn how to use light and shadow in theirs.

In legends, fairies have the ability to make mists. It is possible for people to make mists also. Learning to make mists can help give students an understanding of what scientists do when they attempt to control the weather and stimulate the fall of rain.

PROJECT IDEA

Have students make mists in glass jars using the following instructions.

Materials

Glass jars

Metal trays

Warm water

Ice

Instructions for Making Mists

1. Have students fill a metal tray with ice and let it sit until the tray gets very cold.

2. Then have students fill a glass jar with about one inch of warm water and place the cold tray on top of the jar. A misty cloud will form in the jar when the warm moist air above the water meets the cold air just under the tray of ice.

3. Explain to students they have cooled the air so that it is unable to hold water vapor, and it condenses. This is the way mists are formed.

SUGGESTED READING

Weatherwiz Kids. "Fog." *http://www.weatherwizkids.com/fog.htm.*

> Contains instructions for making fog in a jar and explanations of what occurs.

Web Weather for Kids: Clouds. "Make Fog in a Jar!" *http://www.ucar.edu/educ_ outreach/webweather/cloudact1.html.*

> Contains instructions for making fog in a jar and explanations of what occurs.

TOPIC 3. The Fairy Realm

Legends of fairies appear in many places, but the best-known fairy legends come from western Europe, particularly from Great Britain and Ireland. In Irish folk belief, fog, mist, and sudden gusts of wind are often attributed to fairies. Beneath the fog and within the mist lies a magical world, and fairies can conjure up these phenomena to cover clues that might divulge the location of fairy people to human beings. Fairy places usually have a king or a queen who rule the land, much like Morgan le Fay ruled Avalon. Like human beings, fairies usually live in houses or huts, while fairy royalty live in castles. In legend, fairy castles are made of shimmering gold, silver, crystal, or pearls.

PROJECT IDEA

Have students choose a famous fairy or class of fairies and write a story that explains how these fairies relate to nature. The two Web sites in the Suggested Reading section list numerous types of fairies, many of which have connections to nature.

SUGGESTED READING

Fairies of the Realm. "Types of Fairies Explained." *http://www.usa2076.com/fairies/types. htm.*

A list and description of the different kinds of fairies.

TOPIC 4. Phantom Ships

Mirages on the sea gave rise to numerous legends of phantom ships that sail the seas forever and materialize from time to time in certain areas of the world. Like magical islands, these phantom ships are surrounded by fog and mist, and though they have been spotted by people over centuries each one of these ships is simply a mirage. Usually, phantom ships sail in the absence of wind, an atmospheric condition that is needed for fogs, mists, and mirages to occur. These ships look so real that people have let their imaginations run away with them. In stories, phantom ships are navigated by ghostly crews and are doomed to remain at sea forever and never reach port.

The most famous legend of a phantom ship is the legend of the *Flying Dutchman*, the name given to a ship that is often seen floating above the water in the Cape of Good Hope off the coast of Africa. The *Flying Dutchman* is commanded by Captain Vanderdecken, a criminal who is condemned to sailing the seas forever in penance for his crimes.

PROJECT IDEA

The following Web sites contain legends of the *Flying Dutchman*. Read one of these legends to your students and then discuss it in class. You might want to have students write a legend of their own that explains the appearance of a phantom ship.

SUGGESTED READING

"The Flying Dutchman Legend." *http://ms.essortment.com/dutchmanflying_rrqy.htm.*

"The Flying Dutchman Legend." *http://pirates.itgo.com/whats_new.html.*

TOPIC 5. King Arthur Legends

Legends of King Arthur and the knights of the Round Table are some of the best-known stories of the Middle Ages. These stories were written as history, as if the legendary king and all the characters who populated his stories truly existed. Morgan le Fay was just one of the many characters that populate Arthurian legends. There are many others who are just as colorful. The stories of King Arthur's Camelot have enchanted people for hundreds of years and have become very much a part of the history and legend of Britain.

PROJECT IDEA

Have students choose one of the following topics and write a short essay about some aspect of the Middle Ages. The sources listed in the Suggested Reading section contain information that should help students with their essays.

Topic List

Armor

Camelot

Castles

Chivalry

Excalibur

Feasting

Heraldry

The Holy Grail

Knighthood

Troubadours

SUGGESTED READING

Crossley-Holland, Kevin. *The World of King Arthur and His Court: People, Places, Legends and Lore*. New York: Dutton, 1998.

> Contains stories of King Arthur and his court, including stories of Morgan le Fay, King Lancelot, Guinevere, Merlin, and many other legendary figures. Has excerpts from medieval texts and beautiful illustrations.

Roberts, Jeremy. *King Arthur*. Minneapolis, MN: Lerner Publications Company, 2001.

Examines King Arthur and the knights of the Round Table from both a legendary and a historical perspective. Contains information on many of the characters that populate the Arthur legends.

TOPIC 6. Fighting Pollution and Smog

Smog is a dangerous condition that occurs when fog mixes with smoke and dust and chemical particles. Smog has become a serious problem in cities that have large factories and numerous cars on the highway. Smog forms when layers of air are trapped over a city. A layer of warm air moves over a layer of cold air and acts like a lid that traps the smoke and dust and chemical particles beneath it. Those dangerous chemicals rise into the air and cannot break through the lid, so they remain low to the ground and pollute the cities. Crops die, wood rots, and metals corrode. People have trouble breathing and often develop health problems, such as bronchitis, asthma, and emphysema.

One of the most dangerous smog-producing chemicals is carbon dioxide. Factories that burn coal release large amounts of carbon dioxide into the air. Over time, the release of those chemicals has caused changes in the Earth's climate. The Earth is getting warmer, and the problem is worldwide. Global warming and the greenhouse effect have become serious environmental concerns, and unless we do something about it, the Earth will be damaged irreparably.

PROJECT IDEA

There are numerous ways to teach students about global warming and the greenhouse effect, and educating students about ways to reduce smog is a logical beginning. It makes the problem seem manageable and it stresses the importance of every person in every nation doing their part to prevent the unnecessary release of toxic chemicals into the air. Have students design a pamphlet to educate people about the dangers of smog and what they can do to help. The sources in the Suggested Reading section contain information that should help students complete their work.

SUGGESTED READING

Gutnik, Martin J. *Experiments that Explore the Greenhouse Effect.* Brookfield, CT: Millbrook Press, 1991.

> Describes the movement of air and how air gets polluted. Explains the greenhouse effect and discusses how people's activities affect the Earth's climate.

NOAA. "Greenhouse Effect." *http://www.oar.noaa.gov/k12/html/greenhouse2.html.*

> Explains the greenhouse effect for middle school science students. Has links to activities.

Pringle, Laura P. *Global Warming: The Threat of Earth's Changing Climate.* New York: Seestar Books, 2001.

Discusses Earth's changing climate and weather plans and discusses what we can do to protect the environment.

—— SUGGESTED READING FOR TEACHERS

Ashe, Geoffrey. *Mythology of the British Isles.* Pomfret, VT: Trafalgar Square Publishing, 1990.

Green, Miranda J. *Dictionary of Celtic Myth and Legend.* London: Thames and Hudson, 1992.

> Dictionary of terms in Celtic myth, religion, and legend. Includes Intro and bibliography.

Greenler, Robert. *Rainbows, Halos and Glories.* New York: Cambridge University Press, 1980.

> Explains atmospheric optics topics that involve reflection and refraction and includes computer simulations of optical phenomenon.

Isle of Avalon. *http://www.isleofavalon.co.uk/.*

> Discusses Glastonbury and the Isle of Avalon.

Lynch, David, and William Livingston. *Color and Light in Nature.* New York: Cambridge University Press, 1995.

> Discusses and explains optical phenomena in the atmosphere.

Meinel, Aden, and Marjorie Meinel. *Sunsets, Twilights, and Evening Skies.* New York: Cambridge University Press, 1983.

> Focuses on atmospheric phenomena and particularly on color in the atmosphere.

Minnaert, M. *The Nature of Light and Colour in the Open Air.* Mineola, NY: Dover Books, 1954.

> Discusses a wide variety of optical phenomena and offers explanations for their causes.

NOAA Research. *http://www.oar.noaa.gov/education/.*

> Contains links to various activities on weather and atmospheric science for students and teachers.

State of Tennessee. "Air Quality Lesson Plans and Data." *http://www.tnrcc.state.tx.us/air/monops/lessons/lesson_plans.html.*

> Contains links to numerous lesson plans and activities for grades K–12. Includes links to sites on topics such as air pollution, carbon dioxide, carbon monoxide, the ozone layer, and general meteorology.

"The Superior Mirage: Seeing Beyond." *http://www.islandnet.com/~see/weather/elements/supmrge.htm.*

> Explains the superior mirage and gives detail about the Fata Morgana and other types of superior mirages. Contains diagrams. Also has links to information on inferior mirages.

ThinkQuest. "Fata Morgana." *http://library.thinkquest.org/C003603/english/phenomena/fatamorgana.shtml.*

Gives information about mirages, particularly the Fata Morgana.

Timeless Myths. "Arthurian Women: Morgan le Fay." *http://www.timelessmyths.com/arthurian/women.html#Morgan.*

Gives information about the character and myth of Morgan le Fay.

University of Rochester. "The Camelot Project." *http://www.lib.rochester.edu/camelot/.*

Contains information of the people, places, symbols and motifs in the legends of Camelot and King Arthur.

Wicker, Crystal. "Weather Wiz Kids." *http://www.weatherwizkids.com/fog.htm.*

Contains information on the weather and includes weather folklore, jokes, games, experiments, and links to related sites.

5 .... Storms

—————————— THE MYTHS OF STORMS

In the myths of ancient India, Indra was the god of storms. Along with Surya, the Sun god, and Agni, the Fire god, Indra was one of the strongest powers in the land. He ruled the atmosphere, and he controlled all phenomena believed to arise in the atmosphere, such as the wind, the rain, the thunder, the lightning, and the storm.

Myths of Indra and his companion storm deities, the Maruts, permeate Indian mythology, but one of the best-known myths involves the Storm god's battle with Vritra, the drought demon. At the end of each summer, Vritra kidnapped the cloud cattle and imprisoned them in a cave, thereby keeping all the waters of the world captive. Indra slew Vritra and released the cloud cattle, and by doing so he rose above Surya and Agni and became the highest power in existence. Read the following story and learn how Indra slew the drought demon. Then, identify beliefs about storms that appear in the myth and compare them to the science that arose years later.

"HOW INDRA SAVED THE CLOUD CATTLE," A MYTH FROM INDIA

It is always dry in India in the summer, and by the end of summer it gets so dry that it seems that the land will shrivel up like a gigantic prune or burst into flames from the heat. For months, the storm winds do not blow nor does a drop of rain fall, and there seems to be nothing at all anyone in the land can do about it. Drought plagues the Indian people every year, and every year it seems nearly hopeless. Then, miraculously, relief comes. Relief comes carrying weapons and riding a chariot of gold. A mighty battle ensues in the mountains, and before long, it begins to rain.

"Indra has won the battle," someone always cries, relieved and in awe of the torrential rains that pour from the heavens. "The cattle have been released, and Indra has saved us!" People everywhere look to the heavens and watch as the precious cloud cattle charge across the sky.

The summer battle between Indra and the drought demon has been legend in India for as long as anyone can remember. Thousands of years ago when the land dried up for the first time, Indra was born to help them. This was a time when supernatural beings existed in the world who could do something to control the forces of nature. But it was up to these supernaturals to decide when and how to do it.

There were many supernatural beings at work in India at that time, gods and demons both. At that time, the great god Varuna was the highest god in existence. Varuna had ruled over Earth, sea, and sky, and he controlled all the water of the world—until Indra was born. When Indra was born, it became clear that he alone could save the Earth, and after he proved it for certain by rescuing the cloud cattle, he usurped Varuna's power. Ancient India was at the mercy of deadly forces. Each year, the Indian people nearly starved from the intense heat and long drought. No rain fell on the Earth for months, and with nothing to water the crops everything died.

The year Indra was born India had the worst drought ever. Varuna was powerful, but it took a fierce and mighty warrior to save the Earth when the drought demon captured the cloud cattle. For many months Vritra held the cattle captive in a mountain cave, and he built ninety-nine fortresses to protect them. The people cried for help, and Indra was born to save them. The moment Indra was born he began fortifying himself with a magic elixir called soma. Soma energized Indra, and he quickly grew ferocious and strong. In the dead heat of summer Indra prepared to battle Vritra. The drought demon was the most dreaded demon in existence. He roared so loudly he shook the heavens. But nothing could stop the mighty warrior. Indra drank enough soma for 10,000 cows. With all the strength of the universe behind him, his drive to kill demons soon became the most powerful force in India.

When Indra set out to rescue the cloud cattle he had a powerful weapon and an army of men by his side. The weapon was called Vajra, and it was sharp and swift enough to part mountains and split rocks. The army of men were the Maruts, and there were twenty-seven of them—all handsome and all outrageously strong.

The Maruts dressed in gold from their heads to their toes. They wore golden helmets on their heads, golden breastplates on their chests, and golden bracelets on their wrists and ankles. The Maruts were the sons of Rudra, the god of hurricanes, storms, and winds. Rudra was wild and tempestuous. He howled like a bull and shot arrows of death and disease throughout the land. While Rudra was known for destruction, the Maruts had both good and bad qualities. The Maruts remained fiercely loyal to Indra. They could certainly stir the storm winds and agitate the waters, and they rivaled Indra in their relentless drive to save the Earth.

The day Indra rescued the cloud cattle, he took power of the world completely. This fierce and violent warrior drank so much soma that he became invincible. Without hesitation, he climbed aboard his golden chariot. He grasped Vajra, the mighty thunderbolt, in his right hand. Seeing Indra prepare for battle the Maruts quickly rose to help him. They climbed aboard their chariot and readied themselves for battle. They grabbed their shining axes, arrows, and bows. Then Indra cracked his whip and off they flew. Two tawny horses carried Indra, and three swift-footed deer carried the Maruts. They rode on a whirlwind, split open the mountain, and broke through all ninety-nine fortresses before Vritra knew what hit him. Indra slayed Vritra with his thunderbolt, and the clouds scudded out into the sky. In moments, torrential rains fell on the land.

Indra was a great and benevolent guardian of the Indian people, and his heroic victory over Vritra ensured his power as highest god. Because India suffered so terribly during the summer droughts, it made sense that whoever could release the rains should become lord of the land. It was Indra who came to the people's rescue that year and each year after that. By fighting the demon and rescuing the cloud cattle, Indra nourished the fields and ensured that they stayed fertile. So from then on the people worshipped him. It was a great boon to the country the day that Indra saved the world.

. . . .

"How Indra Saved the Cloud Cattle" was created primarily from accounts of Indra as they appeared in *Indian Mythology* by Veronica Ions (New York: Bedrick Books, 1983, reprint).

THE SCIENCE OF STORMS

Storms come in all shapes and sizes. They have both creative and destructive powers. Like rain, they renew the Earth after long droughts, but they can just as easily destroy the Earth with powerful winds and torrential downpours.

Read the following facts about the different kinds of storms. Then refer to the Topics for Discussion and Projects section to see ideas for student research and study.

....
BELIEF: Storms are enormous in size.

In the myth we just read, Indra drank soma and grew to enormous size. This is a common feature of storm myths. In the chapter on wind, we discussed Otis and Ephialtes, the sons of Poseidon who grew nine inches every month. Typically storms start out small and then pick up speed. Hurricanes, cyclones, typhoons, and tornadoes are all ferocious storms. Cyclones occur in the Indian Ocean, and hurricanes are defined as tropical cyclones that occur in the Atlantic Ocean. Tornadoes occur over land. All of these storms are characterized by high winds, which can feed the storm and make it grow bigger.

The difference between storm winds is simply a matter of scale. Tornadoes are much bigger and can be much more destructive than whirlwinds, and hurricanes are much bigger and can be much more destructive than tornadoes. Cyclones and hurricanes are perhaps the most dangerous storms imaginable, and they can be 400 miles across. When a storm that large slams the coast, it causes high winds, floods, torrential downpours, and mass destruction. Oftentimes, tornadoes spin off from hurricanes.

....
BELIEF: Storms bring powerful winds.

Windstorms are the most powerful storms, and windstorms include cyclonic storms, such as hurricanes, typhoons, cyclones, tornadoes, and waterspouts. Take a look at Figure 5.1, which shows the anatomy of a storm. Cyclonic storms swirl around a central eye, which is the innermost part or the center of the storm. In the Northern Hemisphere cyclonic storms always rotate counterclockwise around the eye, and in the Southern Hemisphere they always rotate clockwise. The eye wall is the area of high winds just outside the eye. This is the most dangerous area of the storm where the most violent activity occurs. The eye of a hurricane is usually fourteen to twenty miles across with low pressure and high temperature. The eye is very calm but very dangerous. Because the eye of the storm is so large, people might experience a long period of calm and believe the storm to be over. Devastation can occur because

FIGURE 5.1 · Anatomy of a Storm

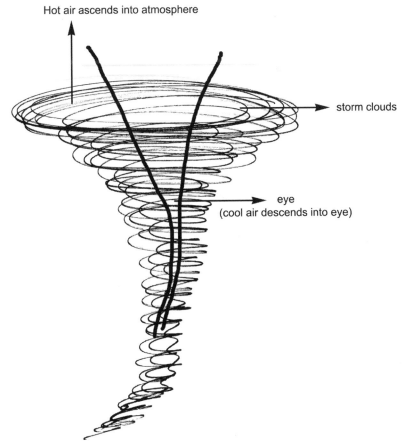

Hot air ascends into atmosphere

storm clouds

eye
(cool air descends into eye)

* spiraling winds and storm clouds move counterclockwise
in the northern hemisphere and clockwise in the southern hemisphere

the storm continues moving, and people fail to recognize that the other part of the storm is still on the way. Hurricanes are perhaps the most dangerous of all cyclonic storms, and they can be extremely unpredictable. Hurricanes can zigzag over the waters or move backward. Sometimes they can stand completely still for a while before sweeping over the waters again.

Tornadoes don't have to stem from hurricanes, but they can. Tornadoes are funnel clouds that extend from the base of a cloud to the ground. They can come from any direction, but usually in the Northern Hemisphere they move from southwest to northeast. They whoosh and they rumble, and extremely strong tornadoes can flatten towns and cities. Tornadoes always occur over land, but a waterspout is a similar phenomenon that occurs over water. Waterspouts extend from the base of a cloud to the water, and they can occur over seas, lakes, bays, and other waters all over the world. Waterspouts are most commonly seen along the southeast coast of the United States, off the coast of Florida and the Florida Keys. Waterspouts can capsize boats and ships. They can also move ashore and turn into tornadoes.

BELIEF: Storms are seasonal powers.

In the United States, most storms, including tornadoes, come in the spring, but most hurricanes come at the end of summer. In the United States, hurricanes pose a serious threat from the beginning of June to the end of November. In India, summer is the season of tropical cyclones and monsoons.

Monsoons occur seasonally when the wind direction shifts and ushers in a change in the weather. Some myths of Indra say that the Storm god rescues the cloud cattle every summer. This is because the Indian monsoon comes every summer. The word *monsoon* comes from an Arabic word that means "season." Every summer, cool, moist winds blow inland from the Indian Ocean and sweep over the land. This is because air always moves from cooler areas to warmer areas. In India at the end of summer, the land heats up and the difference in temperature between the land and the sea increases. The greater the temperature difference, the stronger the monsoons. Monsoon season in India is usually a welcome relief from the drought, but the summer monsoons often bring torrential rains and severe thunderstorms. Sometimes monsoons bring tropical cyclones, and these cyclones can move ashore, can cause flooding, and have at times killed thousands of people. In winter, things are different. At the beginning of winter, the monsoon pattern in India is reversed. The Indian land cools, but the water over the Indian Ocean is still warm. Cool air then moves from the Himalayas and northern India outward, so the winter monsoon brings cool, sunny, and dry weather to India.

BELIEF: Storms bring torrential rains.

In the summer, as the Indian continent heats up, the hot air rises. The cool, moist air over the Indian Ocean moves inland to replace the hot air, and this cool, moist air then rises and condenses to form clouds.

These clouds get trapped inside the Himalayan Mountains, and rain drenches the entire continent.

In the north Atlantic Ocean, hurricanes originate in two zones just north and south of the equator. Hurricanes begin small, as wind whirls softly over the waters, and then the winds pick up speed. Hurricanes sweep across the water, often with tremendous force. Then they move toward the poles. Hurricanes die when they reach the colder areas of the Earth, but in the meantime they can do untold damage. Fed by warm, moist air, these storms spin through the water like a top. They have been known to reach five or six miles in height and travel as fast as 300 to 400 miles a day. People living in coastal areas are particularly susceptible to hurricanes and need to take precautions to protect themselves from these storms during hurricane season. In the north Atlantic, about ten hurricanes form over the waters each year, but usually only one or two of them slam the American coasts.

BELIEF: Storms are benevolent powers.

Myths often expressed conflicts between opposing forces. Sometimes, gods with creative powers fought demons of destruction, but other times the gods with creative powers and the gods with destructive powers were one and the same. Early mythmakers recognized duality in the world; they knew that maintaining a balance between goodness and evil was necessary for survival. A world with too much rain and no drought would destroy the world with flood; a world with only sunshine and no darkness would parch the Earth and reduce it to ashes. Opposition is an essential part of existence, and ancient people knew that fact well because they witnessed it in their world. The alteration of darkness and light turned the seasons, and so did the alteration of rain and storm. For this reason, duality in nature is one of the most fundamental concepts of myth.

The climate of India is one harsh and vicious cycle. Drought and heat are followed by monsoon rains, but after that the land survives. The water that followed drought is both welcomed and feared. When powerful gods of the world possess both good and evil powers, it is because early people knew that natural forces have dual powers. Storm gods are both good and bad because water is both creator and destroyer. Generally, storm gods materialize in the spring to renew the Earth after a long winter. These storm gods release the water that brings the Earth back to life. Vegetation grows after long periods with no food. Clearly, storm gods have creative powers, which means that, often, storm gods acted out of benevolence. Yet people understood the power of the storm, and they knew the ability of these storm gods to cause destruction. Because storms are some of the most powerful forces known, powerful imagery permeates storm myths. Indra and the Maruts carried weapons of destruction—Vajra, Indra's mighty thunderbolt, could part mountains and split rocks, and the axes, arrows, and bows of the Maruts had similar powers. Storm gods in myths all over the world possessed weapons, and these weapons symbolized not only the destructive power of the storm but also its unbridled strength and fertilizing power. Storm gods used their powers to fertilize the Earth, which clearly made these gods benevolent beings responsible for the turn of the seasons and the continuation of life.

BELIEF: Drought is evil and must be conquered.

In the ancient world, drought meant death, and because ancient people feared drought, they injected their myths with drought demons who waged war against storm gods. Oftentimes, myths were told of drought demons who imprisoned the waters. It is these myths that typically reflect a reverence for storm gods. In many lands, people made offerings to the storm gods, believing that these offerings would show the storm gods proper respect and convince them to release the water. When Indra conquered the drought demon, he became the most powerful god in the land.

Droughts can certainly be devastating, though they are a normal feature of climate. Droughts occur almost everywhere in the world, though they vary in severity from place to place. Scientists define drought in numerous ways, but they usually use the term *drought* to refer to a water shortage that occurs when rain fails to fall for an extended period of time. Often drought refers to a dry period that lasts more than a season. In most parts of the world, the amount of rain varies from season to season naturally. In some areas of the world, most of the rain falls in the spring. In other areas of the world, most of the rain falls in the late summer or early fall, the season of hurricanes.

Droughts do not occur every time there is a shortage of rain, but over long periods with no rain the soil dries up. Crops, trees, plants, animals, and people die. Over many months, water levels in streams, rivers, and reservoirs fall, resulting in a severe water shortage. Though drought is a natural feature of climate, it is not entirely a natural phenomenon. People have a large impact on the effects of drought. It's always important to conserve water, but in times of drought especially, people should take care not to stress the Earth's water supply.

— TOPICS FOR DISCUSSION AND PROJECTS

TOPIC 1. Storm Protection

Storm watches or warnings serve as signals to take safety precautions that protect you and your family. Take this opportunity to teach students exactly what these watches and warnings mean and what to do in emergency situations. Tell them that storm watches mean that they should be on alert for possible storm activity, and storm warnings mean that a storm has actually been sighted. Educating students on storm safety means helping them know where to go during a storm, what supplies they might need, and how to reduce the chance of damage to themselves and to property. Hurricanes and tornadoes can be devastating and are frequent threats in many areas of the United States.

PROJECT IDEA

Have students make a brochure that gives instructions for how to remain safe during a hurricane or a tornado. The sources in the Suggested Reading section have much information that will help students create their brochures.

SUGGESTED READING

Alth, Max, and Charlotte Alth. *Disastrous Hurricanes and Tornadoes*. New York: Franklin Watts, 1981.

> Discusses wind force and explains the characteristics of hurricanes and tornadoes. Includes safety information.

American Red Cross. "Are You Ready for a Tornado?" *http://www.disasterrelief.org/ Library/Prepare/tnado.html* and "Are You Ready for a Hurricane?" *http://www. disasterrelief.org/Library/Prepare/hcane.html.*

> Includes information on disaster preparedness including methods of escape and how to assemble disaster supply kits. Gives other information about what devastation these storms can cause and what do when your area receives a storm watch or warning.

TOPIC 2. Making Tornadoes

Anyone who has seen *The Wizard of Oz* knows how a tornado looks when it sweeps across open plains. Strong tornadoes actually can uproot trees and can carry away cows, people, and houses. They don't take them to Oz, though; they carry them long distances and tear them apart or drop them on the ground.

PROJECT IDEA

Have students work together to make a tornado to display in the classroom. You can make a simple tornado in a jar or a box, or you can make a more complex tornado by building a tornado chamber. The following Web sites give instructions for making tornadoes.

SUGGESTED READING

Jamison, Emily. "Emily Jamison's Web Quests." *http://students.gvc.edu/emily.jamison/WEBQUEST/TORNADOES.HTML.*

> Contains information, links, activities, and assessments on the subject of tornadoes.

"Make a Twister in a Bottle." *http://whyfiles.org/013tornado/6.html.*

> Has directions for how to make a tornado in a bottle. Contains bibliography.

National Center for Atmospheric Research. "Make a Tornado." *http://www.ucar.edu/40th/webweather/tornado/tornadoes.htm.*

> Directions on how to make a tornado in a jar.

USA Today's Weather Pages. "A Close Up View of Tornadoes." *http://www.usatoday.com/weather/tg/wtorwhat/wtorwhat.htm.*

> Has information about tornadoes.

TOPIC 3. Mapping the Indian Subcontinent

Ancient India was particularly susceptible to the destructive forces of nature. Heat and drought characterized the climate, and heat and drought characterized the myths. The pervasive heat and drought was what made the Indian people worship water gods. Before Indra rescued the cloud cattle, Varuna ruled supreme. He ruled supreme because he controlled all the waters of the world.

PROJECT IDEA

Have students make a map of the Indian continent. Have them show the Himalayan Mountains, the Ganges River, and all the important physical features of the land. Some students may want to make a weather map of the continent and show the pattern of the tropical cyclones and monsoons. The sources in the Suggested Reading section discuss mapmaking and contain instructions for making different kinds of maps.

SUGGESTED READING

Bramwell, Martyn. *How Maps are Made*. Minneapolis, MN: Lerner, 1998.

> Explains the history of mapmaking and provides instructions for making various types of maps.

Bramwell, Martyn. *Maps in Everyday Life*. Minneapolis, MN: Lerner, 1998.

> Describes different kinds of maps and how they're made.

O'Hare, Tim. *Studying the Weather*. Vero Beach, FL: Rourke Publishers, 2003.

> Discusses the study of the weather and what meteorologists do. Includes information on weather forecasts and how to make weather maps.

TOPIC 4. Dragons as Weather Makers

India and many other Asian lands commonly had myths of gods slaying dragons. The type of battle that arose varied from place to place because the dragons in different world myths embodied threatening forces of nature in different world areas. Ancient people commonly believed dragons to be weather makers, though most myths connect dragons to rain, not drought. In China, dragons had the ability to both withhold the rain and produce it.

PROJECT IDEA

Have students create a storm dragon out of papier-mâché. You can do this as a class project and create one gigantic storm dragon to represent the gigantic impact of severe storms. Display your storm dragon in the classroom on stormy days throughout the year.

SUGGESTED READING

Capp, Gerry. *Great Papier Mache: Masks, Animals, Hats, Furniture.* Petaluma, CA: Search Press, Ltd., 1997.

> Contains instructions for various papier-mâché projects.

Seix, Victoria. *Creating with Papier Mache.* San Diego, CA: Blackbirch Press, 2000.

> Contains instructions for creating projects from papier-mâché.

TOPIC 5. El Niño and La Niña

The weather pattern called El Niño has been blamed for weather-related problems around the world, including the droughts and failed harvests in India. When an El Niño starts in the Pacific Ocean, the waters begin to warm, and the winds shift and blow southeast. Scientists recently discovered what causes an El Niño to begin in the Indian Ocean as well as in the Pacific. The wind moves the warm water toward Australia and takes moisture away from India. In 1982–1983, El Niño prevented the monsoon rains from renewing the land after the drought season in southern India. La Niña also affects the monsoon season in India. Because La Niña generally affects the weather in the opposite way from El Niño, La Niña causes an extremely wet monsoon season in India.

PROJECT IDEA

Have students write reports that explain the effects of El Niño and La Niña. Divide the class into two groups and have the boys cover El Niño and the girls cover La Niña. Following is a list of topics related to these weather patterns that should be included in the report. You might divide the two groups into smaller groups and assign the smaller groups one of these topics. Then have the boys and the girls present their reports to the class.

Aspects of El Niño and La Niña

Effects on the oceans

Effects on the winds

Effects on the seasons

Effects on the rainfall

Effects on storms

Effects on animals

Effects on plants

SUGGESTED READING

Arnold, Caroline. *El Niño: Stormy Weather for People and Wildlife.* New York: Clarion Books, 1998.

Explains the science behind El Niño and La Niña and the effect these weather patterns have on people and animals. Includes statistics, photographs, and a helpful glossary.

National Oceanic and Atmospheric Administration. "El Niño Theme Page." *www.pmel.noaa.gov/toga-tao/el-nino/home.html.*

Contains information about numerous aspects of El Niño and La Niña.

Rose, Sally. *El Niño and La Niña.* New York: Simon Spotlight, 1999.

Contains information on the causes and effects of these weather patterns and discusses the differences between them. Presented by the Weather Channel.

Seibert, Patricia, and Jan Davey Ellis. *Discovering El Niño: How Fable and Fact Together Help Explain the Weather.* Brookfield, CT: Millbrook Press, 1999.

Gives information on the discovery, history, and science of El Niño.

—— SUGGESTED READING FOR TEACHERS

How Stuff Works. "How Hurricanes Work." *http://travel.howstuffworks.com/hurricane.htm.*

> Contains a detailed explanation of the science of hurricanes.

How Stuff Works. "How Tornadoes Work." *http://travel.howstuffworks.com/tornado.htm.*

> Contains a detailed explanation of the science of tornadoes.

Landers-Cauley, Diana, Ellen Mantenfel, and Elaine Berry. "Drought: How Dry I Am." *http://www.sandwich.k12.ma.us/webquest/drought/.*

> Explores the effects of drought and the devastation it can cause.

National Center for Supercomputing Applications. "Weather Here and There." *http://archive.ncsa.uiuc.edu/Edu/RSE/RSEred/WeatherHome.html.*

> Weather unit for K–12 teachers with hands-on experiments and lessons on observing, forecasting, and broadcasting the weather.

National Drought Mitigation Center. "Drought for kids." *http://www.drought.unl.edu/kids.*

> Defines drought and explains its effects. Has numerous links to other information about water shortages and water conservation.

"The Online Tornado FAQ." *http://www.spc.noaa.gov/faq/tornado/.*

> Contains an overview of tornado science with clear explanations to give to students.

6 ·················· Snow, Frost, and Ice

—— THE MYTHS OF SNOW, FROST, AND ICE

Many characters in Norse myths personify the harsh elements of winter, including the frost, the snow, the stormy sea, and the frozen Earth. In northern Europe, the Earth thawed for just a few months each year. Winter dominated the land for nine long months, and early settlers in these lands feared the possibility of freezing to death.

Norse mythology focuses on contrasting forces and is often characterized as the "land of ice and fire." Most of Norse mythology came from Iceland, particularly from two sources: the *Poetic Edda* and the *Prose Edda*. The *Poetic Edda* is a collection of poems that were written in the early Viking period. The *Prose Edda* is a collection of stories written in the thirteenth century by a man named Snorri Sturlusen. The myths in both of these sources focus on the forces of ice and fire, and these myths are full of frost giants and fire giants.

Read the following myth from Scandinavia that tells of the marriage of two frost giantesses. Both of these characters were as radiant as the glistening ice but as cold and unyielding as the harsh northern winters. In "The Marriage of Frost," these giantesses marry gods of the summer Sun, and their unions represent a temporary victory for the Sun, who manages to melt the ice and snow for a short time each year.

"THE MARRIAGE OF FROST," A MYTH FROM SCANDINAVIA

In the midst of a bitter cold northern winter, the giant bird Hraesvelgr sat on his icy perch in the far north. He flapped his enormous wings and sent chilling winds through nine worlds for nine long months each year. The nine worlds that were blasted by Hraesvelgr's winds rested one on the other, and they fell into one of three realms. Midgard was the realm of Earth, Asgard was the realm of the gods, and Jotenheim was the home of the frost giants. The frost giants mingled with the gods from time to time, and sometimes they married them. Like the sparkling snow and glistening ice, frost giantesses were both alluring and deadly. There were enough frost giants in the world to take over the Earth most of the time, but once in a while a giantess relented to the embrace of a god and married him and the Earth became a much more pleasant place.

Two of the best-loved gods of Asgard married frost giantesses. Njord was god of the harvest and the summer sea, and his son Frey was the god of warmth and the summer Sun. Both father and son found the frost giantesses irresistibly enchanting. Frey's marriage to the ice goddess Gerd was one of the greatest victories for the gods of Asgard. Njord's marriage to the snowshoe goddess Skadi was of Skadi's own doing, however, and the marriage did not work out very well.

Skadi was a beautiful frost giantess who lived in Jotenheim in the stormy home of Thrymheim with her father, Thiazi. Hraesvelgr chilled Jotenheim constantly, so Thrymheim was constantly bombarded with blasts of icy wind and hail the size of bullets. One day, Thiazi left Skadi alone at Thrymheim and went on a mission. Skadi feared for her father, for the Norse gods were in constant battle with the giants, and Skadi thought that the gods had kidnapped her father and kept him captive somewhere far away. As time went by, Skadi got more frightened, and after a time she realized that the gods had killed her father. She turned stone cold with anger, and her pale eyes turned to solid ice. Skadi walked through the icy rooms of Thrymheim and contemplated what to do. Finally, after some time, she armed herself with weapons and left for Asgard, the home of the gods.

The world at that time was a complex place. Jotenheim was buried deep beneath the Earth, and Asgard lay far above it. A rainbow bridge extended from Earth to Asgard, and a god named Heimdall guarded the rainbow bridge that led from one world to the next. When Heimdall saw Skadi approaching, he realized that the gods in Asgard were in grave danger. The beauty of the frost giantess, Heimdall knew, was only a guise. Her icy grip could be deadly. Heimdall knew he had to make peace with this frost giantess to save the world, and he asked her what retribution she might take for her father's death.

"I wish to take a husband from Asgard," Skadi told Heimdall. "A handsome husband who is as warm as the summer Sun and as fair as the summer wind."

Heimdall spoke with the gods about this and they agreed. However, there was one condition the gods placed on the agreement: Skadi must choose her husband only by his feet. Skadi entered Asgard, and the gods gathered in a circle. Skadi studied their feet and made her choice. She chose Njord, the god of the harvest and the summer sea. Njord looked weathered from long days of seafaring, and he smelled strongly of salt.

Njord smiled at Skadi warmly, captivated by her glistening beauty. Skadi obliterated his smile immediately with an icy stare. Nevertheless, Njord asked the frost giantess to live with him at his shipyard. Skadi refused. She wished only to live at Thrymheim, she told him, for she thrived in the chilly halls of her father's home. She wanted to spend her life in the bitter cold mountains where the snow was thick and solid and she could ski to her heart's content. Skadi and Njord made an agreement, however, determined to give their marriage a try. They would live half the year at Thrymheim and half at Njord's shipyard. But the agreement didn't last. Njord and Skadi tried this arrangement for awhile, but sooner or later it became evident that frost and ice could never exist in the presence of warmth and sunlight, nor could sunlight and warmth exist in the presence of bitter cold. The marriage of such opposites was simply not possible.

The marriage of Njord and Skadi did not last, but they did have two children while they were together. These children were Frey and Freya, and both of them loved the warmth and the sunlight as much as their father did. Frey and Freya grew up to become two of the best-loved gods in the northern lands. Freya was the goddess of love, and Frey was the god of Sun and rain. Frey assured that the spring showers, the summer warmth, and the bountiful harvests returned each year after the frost giants released their icy grips on the Earth. Frey traveled on a magic ship called *Skidbladnir*, a ship that could sail on land, sea, and sky and that represented the summer clouds. But as much as Frey loved sunshine he fell madly in love with an enchanting frost giantess, just as his father had. Gerd, the object of Frey's affection, was absolutely radiant. Her shimmering eyes and her bright glistening smile lit the Earth and the sky for miles around.

Frey saw the lovely frost giantess one day while he was in the great god Odin's hall looking out on all nine worlds. When he looked north toward Jotenheim, the land of the frost giants, he saw Hraesvelgr, the giant plumed bird sitting on his perch, poised and ready to flap his monstrous wings and send icy winds toward Asgard. Then, in an instant, a glittering light filtered through the air. Frey was taken aback by this light, for it was as captivating a light as he had ever seen. His eyes turned in the direction of the light where he saw the magnificent hall of the giant Gymir. Frey saw a woman walking out of the hall, and she was the most beautiful woman the Sun god had ever seen.

This woman's name was Gerd, and she was Gymir's daughter. She had clothes that sparkled so brightly they sent glistening drops of light dancing in every direction. As Gerd walked outside her father's hall the entire world glowed. When she moved, light flashed across the Earth, sea, and sky and filtered through all nine worlds, turning everything in existence into a winter wonderland.

From the minute Frey laid eyes on Gerd he longed for her, and over time he grew sick with longing. Unsure of what to do, he finally enlisted his servant Skymir to help him win Gerd's favor.

"Go to Jotenheim and visit Hlidskjalf, the home of the giant Gymir," Frey told Skymir. Bring Gymir's daughter to me and bring her to me quickly."

So Skymir set out on a journey to the frozen land of Jotenheim. It was a long, rough journey and a terribly dangerous one. Not only did Skymir face stultifying winds and deadly cold when he reached the frozen land, but before he could get there he had to travel through a curtain of fire.

"Take my horse," Frey had told Skymir. "My horse will break the curtain. He will not flinch from the fire, and he will take you swiftly through the flames."

Frey gave Skymir his trusted horse and a magic sword to fight the giants. Skymir traveled over stone cold land and into the darkness. The horse galloped up a mountain pass. Then, just as Frey had said, he blasted through the curtain of fire. When they emerged from the flames, Skymir and the horse landed in the midst of a cold desolate land surrounded by hills of rock. Quickly Skymir spotted Gymir's hall. A pair of ferocious watchdogs were chained to the gate out front, roaring like monsters and threatening to devour anyone who attempted to enter the home of the frost giant.

From inside Gymir's hall Gerd heard the commotion outside. She inquired of her servant what was the matter.

"There is a visitor outside who wants to enter," the servant told her. "He wishes to see you."

"Invite him to enter then," Gerd said coldly.

The servant ordered the dogs to allow Skymir entrance, and Skymir walked with confidence into the icy rooms to meet the giantess. As soon as Skymir entered, Gerd greeted her visitor. She was dressed all in white and her eyes remained locked in an icy stare. Skymir wasted no time in telling her of Frey's request. Skymir had come prepared with gifts, and he offered Gerd a handful of golden apples in exchange for her love.

"My love cannot be bought," Gerd replied, in a voice stone cold with anger. "I will never submit to Frey's embrace. Never."

Obviously, Gerd's favor would not be won easily, but Skymir was not easily discouraged. He offered the icy giantess a gold ring, but still Gerd resisted.

"I have no interest in such wealth," Gerd told Skymir. "My father is wealthy enough, and I have every intention of remaining here in his hall, with him."

For a while, it seemed as if there was nothing Skymir could do to melt the heart of the frost giantess. Gerd's heart remained frozen for nine long months while Frey grew increasingly impatient. This beautiful goddess continued to glitter and shine, but she remained ice cold and untouchable—locked within Hlidskjalf, the home of her father. Skymir knew that he must break through the ice somehow or he could never return to Asgard. Finally, Skymir resorted to threats. He drew the magic sword that was entrusted to him by Frey, and he told Gerd that she must submit to his demands.

"If I touch you with this magic sword," he told the giantess, "you will be doomed to sit with Hraesvelgr in his nest, blasted by gusts of wind forever more. You will see no one, and you will do nothing—nothing but gaze longingly upon the nine worlds, none of which you will ever enter again."

Gerd could see that the sword Skymir held was magic, and she knew that this servant of Frey had cast a spell on her that she could not resist. Her icy eyes filled with tears. They dripped, and then flowed, and then splattered on the stony floor of her father's hall.

"Tell Frey that I will meet him in the forest nine nights from now," Gerd said. Then she turned away. Skymir bowed, mounted his horse, and rode swiftly and surely back to Asgard to inform Frey of his victory.

Indeed it was a victory, for just as Gerd had promised, she arrived in the forest to meet Frey nine days later. When they met, whether it was the magic of Skymir's spell or the magic of love itself, Gerd was simply powerless. The warmth of Frey's embrace melted her heart as surely as the summer Sun melted the frozen fields, and in no time at all the frost giantess married the Sun god.

. . . .

"The Marriage of Frost" was adapted primarily from two Norse myths, "Skirnir's Journey" and "The Marriage of Njord and Skadi." Both myths appear in *The Norse Myths*, by Kevin Crossley-Holland (New York: Pantheon, 1980).

— THE SCIENCE OF SNOW, FROST, AND ICE

In a land where ice and snow cover the ground much of the year, it seems miraculous when the world thaws. The myth we just read is a metaphor for this thaw. Frey's victory over Gerd represents the Sun's victory over the winter cold. Read the following beliefs that surfaced in "The Marriage of Frost" and compare those beliefs to the science that emerged years later. A list of topics for discussion and projects follows.

BELIEF: Ice and snow glisten and sparkle.

Most people would say that crystals glisten and sparkle. They do, and that's because snowflakes are made of ice crystals. These ice crystals form in nimbostratus clouds. When the temperature in the clouds is below freezing, the water droplets in the clouds freeze to form ice crystals, though these ice crystals are so small they can only be seen through a microscope. These tiny ice crystals build up in the clouds and stick together. Then they turn into snow crystals. Like raindrops, snowflakes fall when the clouds get too heavy. Some snowflakes are made of only a few ice crystals and are extremely small, and other snowflakes are made of many ice crystals and are much larger.

No two snowflakes are exactly alike, but they are all basically hexagonal in shape and can be grouped into general categories. The type of snowflake that forms depends on the temperature and humidity of the cloud from which it originates. In 1951, the International Commission on Snow and Ice identified seven basic types of snow crystals.

BELIEF: Snow covers the Earth like a blanket.

It may sound hard to believe, but snow can actually keep plants and animals warm. It acts like a blanket because it provides insulation from the temperatures above the snow that, with the blowing air, can be much colder than the snow. The wind chill is a combination of wind speed and air temperature. The temperature we read on outdoor thermometers gives us the temperature close to Earth, but it does not tell us how cold we feel when the wind is blowing. We lose heat when we're exposed to cold air, and the faster the wind speed, the faster we lose heat. The wind chill factor helps us understand how much faster we lose heat when the wind blows. The lower the wind chill, the more problems we can experience due to the cold. People can experience frostbite and hypothermia in extremely low wind chills, so in extremely low wind chills we must take extra care to keep ourselves warm outside.

Animals who live in cold weather know how to keep themselves warm outside. Birds grow a thicker layer of feathers to insulate themselves from the cold, and mammals grow a thicker layer of fur. Some animals grow an extra layer of fat to keep warm. Shivering helps animals keep warm too and so does eating. When we shiver, our muscles contract and expand, and this produces heat. Eating helps animals keep warm because food gives animals energy, which the animals convert to heat. Animals that live in cold climates keep themselves warmer than their environment—they must to survive.

Different animals have different ways of surviving the winter. Some animals keep warm by migrating to warmer places. Many types of birds migrate in the winter, but so do some bats, caribou, elk, whales, fish, and insects. Animals that do not migrate must adapt to the cold weather. Some animals burrow under the snow where they know they can keep their temperature constant and remain much warmer than the air. Underneath the snow, animals get heat from the Earth underneath them and from the light above them. Light penetrates the snow. As the area underneath the snow warms, the snow melts and makes space for the animals to live.

BELIEF: Snow falls from the sky.

Like rain, snow falls from the clouds. We learned in the chapter on rain (Chapter 1) that precipitation begins as ice crystals inside clouds. Snow begins the same way. When it rains, the ice crystals melt when they hit the warm air beneath the clouds. When it snows, the air beneath the clouds is freezing. For snow to reach the ground, the air below the clouds must stay at freezing temperatures most of the way to the ground.

Snow falls from the clouds, but it doesn't have to fall from the clouds for us to experience snow on Earth. In a blizzard, snow doesn't have to fall from the sky at all. A blizzard can occur when snow that has been lying on the ground blows for an extended period of time. To qualify as a blizzard, the falling snow must be accompanied by winds gusting at at least thirty-five miles per hour, and the snow must lower the visibility. Usually in blizzards visibility must be less than one quarter mile for at least three hours. Blizzards can be extremely dangerous. Not only does the low visibility cause problems for motorists, but also the wind chill can get shockingly low during a blizzard—often as low as −60 degrees Farenheit. Blizzards can also cause power failures that can affect entire cities.

BELIEF: Snow melts.

Many myths about snow attempt to explain how snow melts. The myth of Frey and Gerd explains how snow melts. Frey melts Gerd's heart, as metaphorically the summer Sun melts the ice and snow. Myths from other lands also explain how this happens. One Navajo myth explains that Coyote, the trickster, made the snow melt. He melted it in the first year of the world because he was thirsty and wanted to drink the water. According to this myth, snow has melted ever since that time.

When the air gets above freezing, snow melts. If a lot of snow is piled on the ground, it may take awhile to melt, but eventually as the air gets warmer the ground gets warmer, too. Color also plays an important role in how fast snow and ice melt. Snow and ice melt fastest on roads and driveways because black and dark colors absorb the Sun's light quickly, while white and light colors reflect the Sun's light. For this same reason snow melts faster around the base of trees than it does on open areas of grass because the dark bark of the trees absorbs the Sunlight.

When snowflakes fall through warm air, they melt, but sometimes they freeze again before they hit the ground. When this happens, the snow turns to sleet. Sleet is partially frozen raindrops. Sleet is a combination of rain and snow that can be extremely dangerous because it can form a layer of ice on the ground.

BELIEF: Ice covers the snow in winter.

Snow can be soft and fluffy, or it can be hard and crunchy. Newly fallen snow is ninety percent air, but as the snow begins to melt it solidifies.

Often snowstorms can turn to ice storms, and layers of ice can form over snow that has already accumulated on the ground. Ice that falls in winter storms is called freezing rain. When freezing rain continues long enough for ice to accumulate on the ground, an ice storm occurs. Ice storms can develop when there is snow, strong wind, and freezing temperatures. The snow turns to ice pellets called sleet first, and then it turns to rain that freezes on impact when it reaches the ground. Freezing rain is different from sleet because sleet is already frozen when it is in the air. Freezing rain freezes after it lands—on trees, houses, or anything else it encounters.

Usually, snow and ice fall in the winter when the weather is cold, but ice can form in warm weather, too. Hail generally falls in the spring and summer. Hail can fall anytime a thunderstorm occurs, though instead of falling in droplets like rain it falls in rock-hard ice pellets. Hail is formed when ice pellets get trapped inside thunderclouds and get bounced around for a while before they fall through the sky. As they bounce, these ice pellets get coated with more ice, and the pellets get so big that they don't melt when they fall through warm air; they remain in their solid state.

Hail forms from raindrops but often contains small particles from the air that were carried into the storm by strong winds. Hail can contain tiny pebbles, pieces of leaves and twigs, and even small insects. The inside of a hailstone looks similar to an onion because it's composed of layers. Large hailstones have layer upon layer of opaque and clear ice. Opaque ice forms from small water droplets that freeze instantly. When ice freezes instantly, it traps air bubbles, and this makes the ice look milky white. Clear ice is formed from large water droplets that take a longer time to freeze than the smaller droplets. In this case, the air bubbles escape, and the ice looks like crystal.

BELIEF: Snow, frost, and ice are both good and deadly.

The Navajo myth where Coyote melted the snow to drink explains one way snow is beneficial. Long ago, people understood that snow could provide nourishment. They also learned to use snow to cool their drinks. Some people brought snow from the mountains for the purpose of cooling drinks, and other people used blocks of ice from glaciers to make iced drinks and ice cream. When snow melted and water flowed down from the mountains naturally, it stimulated new vegetation. Everyone knew that melting snow provided the Earth with water, but they also knew the dangers of snow and ice. Early mythmakers created characters who embodied the snow and who symbolized the dual nature of winter. In Scandinavia, many of the frost giants were both lovely and deadly.

As beautiful and alluring as newly fallen snow can be, snow and ice can cause hazardous road conditions that leave many travelers stranded. Ice can knock down power lines and leave people without heat and electricity. And when ice covers plants, it can cut off their air and suffocate them. Another danger occurs when snow cause avalanches. An avalanche can occur when newly fallen snow piles up on snow that is already lying on the ground. If the new snow fails to stick to the layers of snow beneath it, it can slide down the mountains and bury anyone or anything that is on the lower part of the mountains or on the ground below the mountains when the snow falls.

BELIEF: Snow and Ice cannot live together with sunlight and warmth.

Many of the myths of northern Europe reflect an awareness of opposites. Opposition is the rhythm of nature—day and night, light and darkness, summer and winter, warmth and cold. This alteration between winter and summer is necessary for the world to survive. But in myths from northern Europe particularly these opposing forces were in constant conflict. In the myth we just read, that snow and ice cannot live with sunlight and warmth is the reason the marriage of Njord and Skadi fails.

The frost giantesses in the myth were both stubborn and unyielding. When Skadi got angry and her eyes turned to ice, the winter chill settled on the land. When Gerd refused Frey's gifts, the snow refused to yield to the sunshine. Both myths reveal a recognition of the alternation between summer and winter. In the story of Njord and Skadi, the winter overpowered the summer Sun, and in the story of Frey and Gerd, the ice and snow finally yielded, and warmth returned to the land.

In Iceland, where these myths originated, the snow and ice do exist when the Sun shines. This is because the snow and ice have become so thick that they never melt completely. In places much warmer than Iceland, snow often remains on mountaintops, even when summer returns to the lowlands. In extremely cold areas, glaciers form. In Iceland large masses of ice remain year-round, so these glaciers formed long ago. Glaciers are made of ice, but they began as snowflakes. The snow piles up on the land and never melts away because more snow continues to fall on top of it. The snow melts and freezes, and more snow melts and freezes, and in time it forms a huge mass of ice. How fast the ice forms depends on the temperature of the air.

As the snow continues to fall and melt, the ice gets extremely heavy, and it begins to move. When this ice forms on mountains, gravity pushes the ice downward, and it slides toward the ground. Rocks fall from the mountains and get stuck in the ice, and, over time, the rocks erode and form sediment. Glaciers are composed of snow, ice, rock, and sediment. As the ice melts, it flows downward and refreezes. Sometimes it flows all the way to the sea and forms fjords and icebergs.

— TOPICS FOR DISCUSSION AND PROJECTS

TOPIC 1. The Anatomy of Snowflakes

All snowflakes are basically hexagons, but they take millions of shapes. Some snowflakes look like the branches of trees. Other snowflakes are much more complex. In 1951, the International Commission on Snow and Ice created a classification system for snowflakes, which identifies seven different types of snow crystals. Each of these types of snow crystals forms under different weather conditions. Star-shaped snowflakes form when it is 0–20 degrees outside, for instance. You can read more about the different kinds of snowflakes and how they form on the Web sites listed in the Suggested Reading section.

PROJECT IDEA

Have students make snowflakes to hang in your classroom. All you need for this project is white paper and scissors.

If you are looking for a more complicated project that you can do as a class, grow your own snowflakes. The complete instructions for this project are available on Caltech's Web site "Guide to Snowflakes," listed in the Suggested Reading section. This project allows students to closely examine snow crystals. It is a good project for students who live in areas of the country where it never snows. Caltech's Web site is an excellent resource for teachers who want to gain an understanding of the physics and chemistry of snow.

SUGGESTED READING

Caltech. "Guide to Snowflakes." *http://www.its.caltech.edu/~atomic/snowcrystals/class/class.htm*

> Has lots of information on snowflakes and snow crystals. The "Guide to Snowflakes" explains the anatomy of snowflakes and contains lots of pictures to give students ideas for creating snowflakes for the classroom. Includes classification chart of the types of snowflakes and numerous activities, especially for students who live in areas where they can get their hands on real snow.

StarrySkies.com. "The Science of Snowflakes." *http://starryskies.com/articles/dln/12-99/snowflakes.html*.

> Explains the science of snowflakes and how they form.

TOPIC 2. Giants in Scandinavian Myth

Giants appeared in Scandinavian myths as colossal beings with super-human powers. They symbolized the natural forces that characterized the northern landscape, such as frost, ice, snow, and hail. Early settlers in the northern lands fought a constant battle with the cold, so in myths the Norse gods fought a constant battle with the giants. In Scandinavian myths, frost giants were called Jotens. They lived in the frozen land of Jotenheim, and they thrived in darkness and fog. These frost giants were hostile and destructive. They had bodies made of ice that formed from the frozen land. In Iceland, giants symbolized the most threatening forces of nature, while elves and dwarves embodied the less-threatening forces.

PROJECT IDEA

Have students write poems about a certain group of supernatural be-ings. Students might choose creatures such as giants, elves, dwarves, fairies, gnomes, or any other type of being they like, and they should focus their poems on the characteristics of these beings and on the nat-ural landforms they might represent. The books in the Suggested Reading Section contain poems of supernatural beings and should stimulate ideas for student work.

SUGGESTED READING

Evans, Dilys. *Fairies, Trolls, and Goblins Galore: Poems About Fantastic Creatures*. New York: Simon & Schuster, 2000.

Hopkins, Lee Bennett. *Elves, Fairies and Gnomes*. New York, Knopf, 1980.

TOPIC 3. The Chemistry of Ice and Snow

PROJECT IDEA

Make ice candles in your classroom. This is an activity that requires adult help and supervision but that can be done easily as a class if you have access to a hot plate or a small electric burner. Ice candles are made from ice and colored wax. When the hot wax is poured over ice, the ice melts and the wax hardens. This illustrates the chemical changes that take place when ice melts and when ice impacts warmer surfaces.

Materials

Small containers to hold the candles, one container for each student (Milk cartons from the school cafeteria work well.)

Popsicle sticks, one for each student

Parrafin wax, which comes in large blocks (One block of wax should be sufficient for every five students.)

Crayons

Masking tape

A ball of string

Wax paper

Ice cubes

A hot plate or an electric burner

A pot with a spout

How to Make Ice Candles

1. *Make the wicks.* Unwind the ball of string. Place a block of wax in the pot and heat it on the hot plate or electric burner. Dip the string in the wax. Remove the string and let it cool and harden on a piece of wax paper. When it hardens, cut the string into pieces, one for each student. The pieces should be long enough to extend from the bottom of the milk cartons to a couple of inches above them.

2. *Make the containers.* Have students cut the tops off of the milk cartons. Then have them tape the wicks to the inside bottom of the milk cartons, in the middle. Students will use the popsicle sticks to hold the wicks in place. Have them lay their popsicle sticks across the tops of the cartons and wrap the wicks around the sticks. Tell them that their wicks should stick up straight in the middle of the milk cartons.

3. *Make the candles.* Decide what color you want to make the candles and then shave or cut two or three crayons of that color into small pieces. Melt the wax and add

the crayon cuttings to the wax. Then have students fill their milk carton about two-thirds of the way with ice cubes. Pour the hot wax on top of the ice in each of their containers. Make sure all of the ice is covered.

4. *Discuss what happens.* When the hot wax is poured over the ice cubes, the wax will harden and the ice will melt. Tell students that this is what happens when warm air and sunlight penetrate the ice. The next day, have students peel off the milk cartons from their candles. Water will pour out of holes created by the ice. Explain to students that this is what happens when snow melts.

TOPIC 4. Ice Age Animals

Ice ages are times when large areas of the Earth were covered with glacier ice. In the last 2 million years, the Earth has had twenty ice ages, but when people refer to the Ice Age as if it were a singular time period they usually mean the last time that North America, Europe, and Asia were largely covered with glacial ice. This Ice Age began 2.5 million years ago. During this time, numerous kinds of large animals roamed the Earth. These giant animals were giant forces, just as the ice itself was a giant force at this time and in Scandinavian myths.

PROJECT IDEA

Have students write reports about animals of the Ice Age, such as the wooly mammoth, the mastodon, or the saber-toothed tiger. The sources in the Suggested Reading section contain information about Ice Age animals that should help students with their work. There may be many more sources on Ice Age animals in your school library.

SUGGESTED READING

Denver Museum of Natural History. "Ice Age Giants." *http://www.dmnh.org/iceage/ia_giants/index.html.*

Enchanted Learning. "Ice Age Mammals." *http://www.zoomdinosaurs.com/subjects/mammals/Iceagemammals.shtml.*

Hehner, Barbara. *Ice Age Cave Bear: The Giant Beast That Terrified Ancient Humans.* New York: Crown, 2002.

————. *Ice Age Mammoth: Will This Ancient Giant Come Back to Life?* New York: Crown, 2001.

————. *Ice Age Sabertooth: The Most Ferocious Cat That Ever Lived.* New York: Crown, 2002.

TOPIC 5. The Land of Ice and Fire

Many of the Norse myths were written in Iceland, which has been called the land of ice and fire. In "The Marriage of Frost," Skymir had to travel through a curtain of fire to get to Jotenheim, the land of frost and ice. In Iceland, glaciers and volcanoes exist side by side. Ice caps are one form of glacier, and ice caps often cover volcanoes. Iceland has many different kinds of glaciers, not just ice caps. During the Ice Age, glaciers covered almost all of Iceland, and today they cover about ten percent of it. Iceland has valley glaciers, which are the smallest type of glacier. They are also called alpine glaciers, and they remain in mountain valleys. The largest glaciers are called ice sheets. Ice sheets cover entire continents. There are only two ice sheets in the world, one in Antarctica and one in Greenland. The ice sheet in Antarctica is the largest; it covers 5 million square miles. The ice sheet in Greenland covers 700,000 square miles. Due to global warming, all the world's glaciers are now shrinking.

PROJECT IDEA

Discuss the effects of global warming on the world's glaciers.

SUGGESTED READING

Environmental Protection Agency. "Global Warming Kids Site." *http://www.epa.gov/globalwarming/kids/index.html.*

> Contains a detailed explanation of global warming and its effect on climate.

TOPIC 6. Winter Animals

Animals who are native to cold climates have ways of protecting themselves from freezing. The pets we keep do not have these natural defenses against cold weather, however, so as responsible pet owners we have to protect our animals against the cold.

The Humane Society issued a press release with recommendations for keeping all pets safe in the winter. The press release appears on the Humane Society's Web site, which is printed in the Suggested Reading section. Use this lesson on ice and snow to discuss the Humane Society's recommendations with students. According to the Humane Society, you should keep pets indoors when the temperature drops below freezing. Remember that the wind chill makes the temperature much colder, and even when the temperature is above freezing you need to protect outdoor dogs from the wind. Make sure the dog has a doghouse and that it is large enough for the animal to be comfortable but small enough to contain the animal's body heat. The doghouse should be dry and free of drafts. Hang a piece of heavy plastic or some other waterproof cover over the door to the doghouse. Face the doghouse away from the wind. Cedar shavings or straw make good coverings for the floor, and the covering should be at least a couple of inches thick. The Humane Society also explains that animals need more food in winter and that you should make sure they have water that is not frozen. Plastic bowls are best for outdoor dogs because in cold weather the dogs' tongues can stick to metal bowls.

PROJECT IDEA

Review the following list of animals that live in cold weather. Have each student choose an animal from the list and write a report about the animal. In their reports, students should include information about the animal's habitat and how it keeps itself warm in cold temperatures.

Seal	Polar bear	Lemming
Walrus	Puffin	Ptarmigan
Fox	Reindeer	Musk ox
Whale	Snow goose	Moose
Arctic fox	Snowy owl	Seagulls
Ermine	Arctic hare	Wolverine
Arctic tern	Caribou	Wolf
Bear	Sheep	Shark

SUGGESTED READING

Enchanted Learning. "Arctic Animals." *http://www.enchantedlearning.com/coloring/arcticanimals.shtml.*

Contains an extensive list of Arctic animals and includes information about their habitats and survival techniques.

Hopping Egan, Lorraine. *Interactive Polar Regions: Cool Activities, Projects, Games Maps, and the Latest Information to Help You Explore these Icy Regions.* New York: Scholastic, 1999.

Contains numerous science experiments and other projects dealing with the science, history, and culture of the polar regions. Includes a card game about animal survival techniques.

The Open Door Web Site. "Activity in a Changing Climate: How Warm-Blooded Animals Stay Warm." *http://www.saburchill.com/ans02/chapters/chap030.html.*

Contains lots of information about the Arctic climate and the different ways animals keep themselves warm and safe in cold temperatures.

Thinkquest. "Animals of the Arctic." *http://library.thinkquest.org/3500/.*

Contains stories, pictures, facts, and activities about Arctic animals.

SUGGESTED READING FOR TEACHERS

Alaska Satellite Facility, University of Alaska. "When Glaciers Form." *http://www. asf.alaska.edu:2222/how_form/glacier_last.html.*

Contains lots of great information about glaciers and includes a teachers' guide.

Heidorn, Keith. "Diamond Dust: Snow Without Clouds." *http://www.suite101.com/ welcome.cfm/science_sky.*

Recent articles and lots of other information about snow and weather.

Heidorn, Keith C. "Weather Phenomena and Elements: The Weather Doctor: Exploring the Science and Poetry of Weather and Atmosphere." *http://www.islandnet. com/~see/weather/doctor.htm.*

This site is an incredible resource for learning the basics of atmospheric science. Easily accessible topics on numerous areas of weather phenomena, with lots of helpful links. The writing is rather advanced for young students, but it's great for teachers searching for simple ways to explain the weather to students.

National Snow and Ice Data Center. "The Cryosphere." *http://nsidc.org/cryosphere/ index.html.*

Discusses snow and related phenomena and has a lot of information about glaciers.

7 .... Rainbows

THE MYTHS OF RAINBOWS

People of past times knew that rainbows appeared in connection with rain, so early mythmakers considered rainbows weather portents. Many myths featured the rainbow as a living spirit who could control the water. Some rainbow spirits had the ability to produce the rain, and other rainbow spirits had the ability to stop it. These beliefs led early sky watchers to use rainbows to predict rain and storms.

Rainbow myths were particularly prevalent in Australia, South America, and other lands where agricultural people depended on long rainy seasons to grow their crops. In these areas, the belief that the rainbow controlled the rain made rainbow spirits fertility gods and creative forces. But people who believed that the rainbow stopped the rain and caused drought knew these spirits could be highly destructive as well. Many groups of people believed that the rainbow could swallow all the water of the world, so they feared that the rainbow could swallow people, too. The rainbow god was often some kind of water serpent, sometimes a dragon, sometimes a crocodile, but most commonly a snake. The mythological rainbow snake is widespread in many lands. The notion stems from the dual nature of the snake as a land dweller and a water dweller and from its dual nature as creator and destroyer. Snakes can kill people and animals, often by swallowing them whole. They also have an amazing ability to shed their skin and renew themselves.

Rainbows have captivated people for centuries. Norse myths tell of a rainbow bridge called Bifrost that linked Asgard, the home of the gods, with Midgard, the home of mortals. The rainbow bridge appears briefly in "The Marriage of Frost," the legend that appeared in Chapter 6 of this book to discuss frost, snow, and ice. The rainbow bridge appeared in places other than Scandinavia, however. In Greek myths, Iris was the goddess of the rainbow. She flew along the rainbow bridge and carried water from the Earth to the clouds.

Read the following rainbow legend from Australia. The rainbow snake appeared in legends from all over Australia, and the story is based most closely on a legend that came from Arnhem Land. In Australia, the rainbow snake had tremendous powers. This creature lived in water pools, which made the pools highly taboo. The legend tells what happened when two women discovered such a water hole and disturbed the snake's waters. In many societies, shamans can enter the snake's water pools without disrupting the snake, but no one else can do this. Shamans have the power to make rain, and in this sense they share the snake's powers. When the shaman enters the water hole, the rainbow snake teaches the shaman how to control the rain, and the shaman takes what he learns back to his people.

"THE RAINBOW SNAKE," A MYTH FROM AUSTRALIA

Somewhere between now and then and sleep and awake, there is a place called Dreamtime. Dreamtime holds supernatural beings of all sorts, beings who exist in some imaginary place deep in the recesses of our minds. The Aborigines of Australia know Dreamtime well, and they're familiar with many of the beings who live there. Some shamans and medicine men of different tribes recognize the supernatural beings of Dreamtime in the world today, and some even pay visits to these beings in some sacred realm that ordinary people can't reach.

In Dreamtime, energy flows from every corner of the universe, and the shamans know how to tap it. This energy infuses the Earth with spirit that only some people can understand. People today who hear legends of Dreamtime might come to believe that events that take place there occur in miraculous ways. In Australia, people hear of the Dreamtime often, and people in tune with the land learn to recognize miracles in ordinary places.

A long time ago, mythological beings roamed the land during Dreamtime, and as they roamed, they performed important tasks that shaped the way the world looks today. The formation of the world is what Dreamtime is all about. Dreamtime beings wandered the Earth and created natural landforms, such as rocks and rivers, lakes and streams. Each of these wanderers played a role in creating the world, and each of them continues to exist in one form or another. Both people and animals took mythological journeys in this magical realm, and they shared the ability to create and destroy and to mold the world into a sacred and beautiful place.

Most everyone who sees a rainbow arching through the sky after a summer storm will recognize the world as a place of beauty, but only those who look beyond the rainbow and into Dreamtime understand the miracle of this colored arc. Those who do look into Dreamtime know the rainbow as a magic snake. This snake has the ability to hold all the waters of the world captive or to release them into the sky at will.

Many people who know the ways of the snake understand that this magic snake controls rain and flood. This is because in Dreamtime it was the snake's responsibility to create all the bodies of water in the world, from the deepest oceans to the shallowest pools. Shamans today visit the rainbow snake by retreating into water holes. This is where the snake lives today. The rainbow snake remains coiled up in the water hole during the dry season, holding the water inside the Earth. When it is time to release the water, the snake emerges from the water hole and stands up on its tail. When it reaches into the sky, the rains begin to fall.

People who have heard legends about Dreamtime know that the rainbow snake has tremendous powers. They also know that this serpent god does not like to be disturbed while it rests underground in its water hole. There is a legend about this rainbow snake that involves two sisters, who were also beings of Dreamtime. These sisters journeyed all over the land, and they found the rainbow snake purely by accident. The snake was burrowed deep within a water hole where the sisters stopped to rest, and the snake was not at all pleased that the sisters saw fit to tamper with its waters.

These two sisters, Boaliri and Garangal, were strong and brave and quite capable of taking care of themselves under ordinary circumstances. When they encountered the snake, however, they had trouble. Boaliri and Garangal had been traveling a long time by the time they reached the water hole. They traveled on foot, with Boaliri's young son, and they survived by gathering yams and berries and by hunting possum and kangaroo and bandicoot that they killed with stone spears. It had been a long, rough journey, and when the women reached the water hole, they were tired. They decided to stop for awhile and rest by the water before continuing on their way. Boaliri and Garangal set up camp by the water hole. They built a shelter from the bark of a stringybark tree that grew by the water, and they built a fire and prepared to cook their food. As resourceful and wise as these women were, however, they did not know all there was to know about the forces that powered the Earth and the waters. They did not know that the water hole was taboo or that it was inhabited by a great rock python named Yulunggul. Yulunggul became frightfully angry when anyone came close to his home and disturbed his waters.

Anyone who knows the legends knows that strange things happened in Dreamtime, and people have been trying to explain those things for centuries. In this case, when the two women attempted to cook the animals they had killed, the animals came back to life as they were placed on the fire. One by one, they jumped into the water hole. Now the people who have tried to explain strange things will tell you that when magic like this happened it was the spirits' way of warning the people. Yulunggul got frightfully angry. He blew water into the air and formed great, dark storm clouds.

Boaliri and Garangal, still in shock from watching the animals jump from the fire, now found themselves in the midst of a raging storm. Lightning crashed and thunder boomed. In minutes, rain poured from the sky. The sisters were frightened, and they took cover inside their shelter. Meanwhile, Yulunggul sprung up from the depths of the water hole. He coiled himself around the hut. The two women huddled together and remained perfectly quiet, but it was hard for them to remain calm when the fiery eyes of a snake flashed menacingly through a crack in the door. Boaliri and Garangal knew they had to fight for their lives, so they began to sing and dance for Yulunggul. Fortunately, their songs entranced the snake, and in a short while he retreated into his water hole and fell asleep. In moments, a beautiful rainbow appeared in the sky. It looked like a python of many colors.

I would like to say that Boaliri and Garangal slept peacefully that night and continued on their journey the next day without trouble. But unfortunately that wasn't the case. The rainbow snake retreated for a time, but when he awakened from sleep he remembered that his waters had been disrupted. There are many ways that the legend continues from there, and in some of those legends the snake devours Boaliri and Garangal. But for the purpose of this story, I will say that the snake did not get the better of Boaliri and Garangal, and they did continue on. They continue on today, in that mystical realm called Dreamtime. In Dreamtime, they dance and sing for the snake to encourage the rains. And people who know that they do this often sing and dance with them. When the people sing, it brings fertility to the Earth, and as they dance, the Earth returns to fruitfulness. Yulunggul was greedy, and he kept all the waters of the world locked up inside the water hole. When he was incited to anger, he spit clouds into the sky and released torrential rains on the Earth. Some say that in Dreamtime these rains flooded the Earth, and the Earth began anew after that. Some say that the snake was a great creator god.

It is said that the rainbow snake created all the waterways of the world and that he continues to release the rains when the Earth suffers from drought. The rainbow snake can be both good and malevolent, both creative and destructive. The Aborigines both feared and respected this creature, just as they did the beautiful and mysterious rainbow that arched out of the water hole and stretched up to the sky.

. . . .

"The Rainbow Snake" was created from accounts of this creature and the legends of Dreamtime that appeared on various Web sites and in *The Speaking Land: Myth and Story in Aboriginal Australia* by Ronald M. Berndt and Catherine H. Berndt (Rochester, NY: Inner Traditions Intl. Ltd., 1994).

THE SCIENCE OF RAINBOWS

Myths and legends formed around rainbows because people noticed when and where rainbows appeared. Many times rainbows appeared during departing thunderstorms, so myths credited the rainbow snake for stopping the rain. Societies who fear the rainbow consider this evidence of the arc's destructive powers. Societies who consider the rainbow solely benevolent credit the rainbow snake with ending the rains sent by enemies.

It is curious why so many cultures connected the snake with the rainbow, but they did in part because both a snake and the arc of a rainbow are curved and thin. When the rainbow appears in the sky, some people envision a gigantic snake standing up on its tail. Some say that this snake can drink water from the Earth during the dry season and release it as rain during the wet season. Others say the rainbow snake emerges from the Earth to drink from the sky. In some legends, the rainbow snake can change size. Like the rainbow, this snake can be small or enormous. Like the rain, the snake was considered both creative and destructive. The Aborigines of Arnhem Land knew the destruction that flood can cause, and they believed that the rainbow snake caused the flooding. In some legends, the rainbow snake creates the monsoon rains that end the dry season. These rains rejuvenate the land and restore the land's fertility, but they can also cause coastal flooding and mass destruction. In this area of the world, coastal flooding led to the end of the last ice age.

. . . .

BELIEF: Rainbows are supernatural occurrences.

No one can climb rainbows, and no one can touch them. Some people think rainbows look too beautiful to be real. These colorful arcs might seem supernatural, but they're simply made of sunlight and water. In nature, rainbows arc through the sky when sunlight from the sky hits rain that falls through the Earth's atmosphere. Rainbows occur most often in the summer because this is the most likely time that rain and sunlight will be present at the same time.

For centuries, people have tried to decipher rainbows. They have tried to touch them, to climb them, and to walk underneath them. They have even tried to follow them to the end and reach a pot of gold. You can't touch sunlight, however, and you can't walk beneath the arc. This is because rainbows always appear in raindrops that fall in front of you and they only appear when the Sun is shining behind you. Sometimes rainbows appear far off in the distance, and sometimes they're just a few feet away.

People recognized rainbows long ago, and they told myths to explain their existence. The rainbow got its name because people long ago knew it appeared during rain showers. Some people thought the arc looked like the bows they used for shooting arrows. Superstitions and beliefs about rainbows are numerous, and they vary from place to place. Many of these superstitions have no basis in fact; they are simply based on individual notions colored by cultural beliefs. It's true that everyone has a different perception of the rainbow, however, because no one sees exactly the same rainbow. People do not occupy the same place at the same time, so they see different droplets of water. The rainbows people see depend on which droplets they see because a different rainbow appears in each one.

BELIEF: Rainbows extend from Earth to sky.

Rainbows are actually circles, though from our position on Earth we can't see the bottom part of the circle because it's below the horizon. At sunrise or sunset, we can sometimes see a complete semicircle though. This makes the rainbow look like it extends from the Earth to the sky. The only time we can see the complete circle of the rainbow is when we are well above the Earth. We might see a circular rainbow from an airplane, for instance.

Rainbows are circular because they are formed when light bends. Each color of the rainbow is bent differently. Red bends the least, and violet bends the most. Look at Figure 7.1. This figure shows how light is bent when it passes through water and air.

FIGURE 7.1 · The Bending of Light

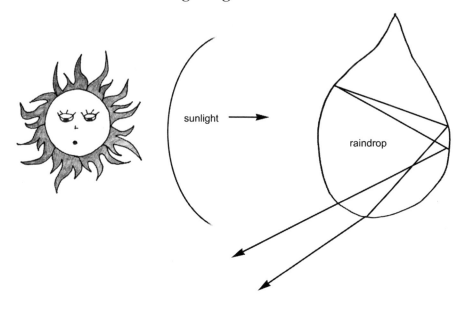

· ·
BELIEF: Rainbows are composed of colors.

Typical rainbows are composed of seven colors—red, orange, yellow, green, blue, indigo, and violet. There are various ways you can remember these colors and their sequence. One way is to remember the name Roy G. Biv. This is the name that has often been given to the rainbow because each of the letters in Roy G. Biv correspond to the beginning letter of the rainbow's colors. Because rainbows seem supernatural, they have been shrouded in superstition for centuries. The Iranian Moslems were one group of people who assigned significance to the colors of the rainbow; red meant war, green meant abundance, and yellow meant death.

Typical rainbows are composed of colors, but the sharpness of the color can vary considerably. This is because rainbows can occur under different atmospheric conditions. They can occur in rather large drops of water or in very small drops of water. Rainbows with the brightest colors occur in large drops of water. This means that the most colorful rainbows occur during thunderstorms. Rainbows with very faint colors occur in small drops of water, so they appear in fogs or mists. These faintly colored rainbows are called fogbows or white rainbows. Fogbows are pale with only slightly tinted borders. They usually appear when street lamps or headlights penetrate the fog and mist. Sometimes at sunset, all the colors of the rainbow except red fade for just a few minutes. In this case, the only thing left is an entirely red rainbow, and red rainbows can be amazingly bright.

BELIEF: Rainbows appear in the sky.

Dewbows are formed by dewdrops, and they appear on the grass. It is often possible to see dewbows in the morning when sunlight shines on the dew. Rainbows can be seen in icicles that hang on trees also. It is possible to see many different kinds of rainbows in the natural world: to see any one of them there must be water in one form or another and sunlight.

There are many kinds of bows that occur in the atmosphere. There are white rainbows and double rainbows. There are fogbows and moonbows. There are also cloudbows. Lunar rainbows, or moonbows, are another form of arc that appears in the atmosphere. A lunar rainbow is formed in much the same way as an ordinary rainbow, but the light comes from the Moon rather than the Sun. Moonbows have color, but the colors are difficult to detect. Moonbows are usually very faint and can only be seen when the moon is full.

Cloudbows look very different from typical rainbows. The water droplets in clouds are much smaller than the water droplets in rain, and this difference in size explains why cloudbows are virtually absent of color. It also explains why cloudbows are much wider than rainbows. Like fogbows, cloudbows are often called white rainbows. Against the white background of the cloud, observers can make out a curved band that is brighter than the cloud.

Rainbows do not always appear as a single arc; there are double rainbows and, theoretically, even triple rainbows. In a double rainbow, the Sun's light is reflected twice before it leaves the raindrop. This makes the second rainbow fainter than the first because light is lost with each reflection. It also reverses the colors in the second arc; red is on the inside, and violet is on the outside. In a triple rainbow, called a tertiary rainbow, the light is reflected three times before leaving the raindrop. In this case, the colors are so faint that tertiary rainbows are rarely if ever seen.

BELIEF: Rainbows reside in water pools.

There are a few other kinds of rainbows that can be found in nature, and these rainbows appear in water pools. A reflection rainbow forms when light rays reflect off a smooth water surface. If there are waves on the water, the bow can look vertical. A reflected rainbow can sometimes be seen in a pond or a large pool of water, and it looks upside down. A reflected rainbow looks flatter than a traditional rainbow.

The rainbow snake in the Australian myth lived in water pools. He stayed there during the dry season, and then he sucked up the water and released it into the air during the wet season. In the myths, rainbow snakes moved from water pool to water pool, and people saw them arching through the air as they traveled. Snakes move quickly and so do rainbows. Sometimes rainbows last only a few minutes before they disappear.

BELIEF: Rainbows have existed since the beginning of time.

Rainbows have existed for as long as the Sun has shone from the heavens and as long as the rain has fertilized the Earth. The rainbow snake in Australian myths existed in Dreamtime, and to the Aborigines Dreamtime is both a legendary time and a state of mind. It was a time when supernatural people and animals created all the beings and landforms, and it was a state of mind that encompasses the intimate connection the Aborigines have with the Earth and the recognition of the sacredness of nature.

— TOPICS FOR DISCUSSION AND PROJECTS

TOPIC 1. Oral History

Many of the legends from early Australia were never written down, or at least not until they had been told for generations. Oral history is just as important a part of history as written history, but unfortunately much of it is lost. Storytelling is an important part of the educational experience for children in Australia. Myths of Dreamtime help explain how the land was formed and how the plants and the animals came to exist.

PROJECT IDEA

Discuss how myths and legends are transmitted from generation to generation. Then have students tell stories aloud to the class. They might want to tell a story that explains how the rainbow formed, or they might want to tell a story that explains what the rainbow snake did to form the land. Students can review some of the superstitions about rainbows to get ideas for their legends. The sources in the Suggested Reading section include legends of Australian Dreamtime and books that contain beliefs and superstitions about rainbows. Students can base their stories on one of the rainbow superstitions and create a legend that has a similar feel to other Australian myths and legends that they read.

SUGGESTED READING

Australian Museum. "Indigenous Australia." *http://www.dreamtime.net.au/dreaming/storylist.htm.*

> This section of the Australian Museum Online contains selected stories of Dreamtime. The site covers indigenous Australia and includes much useful information on Dreamtime and other aspects of Australian culture and spirituality. Has numerous links and curriculum guidelines for teachers and facts for kids.

Rahoorkhuit Network. "Eingana." *http://www.rahoorkhuit.net/goddess/goddess_quest/eingana.html.*

> Has Australian legends, activities, and information on Australian art and culture. Has lots of information on the rainbow snake.

State Library of Queensland. "Aboriginal Resources for Young People." *http://www.slq.qld.gov.au/pub/festivals/aboriginal/young.htm.*

Contains resources on Aboriginal Australia for kids and teachers. Has booklists and links to Dreamtime stories and Web sites containing other cultural info on Dreamtime.

Thinkquest. "The World of Aborigenes." *http://library.thinkquest.org/C005486/homeeng.htm.*

Thinkquest site that explores Aboriginal Dreamtime.

TOPIC 2. Rainbow Symbolism

Advertisers often capitalize on the rainbow's appeal by using rainbows to sell products. Rainbows symbolize goodness and promise. They symbolize treasure, too, as is the case in stories that speak of following the rainbow to a pot of gold. Advertisers also recognize symbolism in the colors of the rainbow.

PROJECT IDEA

Initiate a discussion on the use of rainbows in advertising to lead into the power of advertising in general. You might want to discuss the use of rainbows to sell specific products, such as Rainbow Bread, Lucky Charms, *Reading Rainbow*, and Rainbow Book Company. Ask students what message these companies want to convey by using rainbows in their ads. Then have students create a product and an advertisement of their own. Have them create a billboard or a poster for their product and write a slogan that contains symbolism associated with the colors of the rainbow or the rainbow itself. They might want to use a rainbow to advertise paint, for instance, to capitalize on the colors of the rainbow. Tell students to use their imagination. They can decorate their billboards or posters with rainbows and hang them in the classroom. The books in the Suggested Reading section contain information on the use of advertising and the impact advertising has on young people. These books provide plenty of ideas for taking class discussions on advertising in different directions.

SUGGESTED READING

Day, Nancy. *Advertising: Information or Manipulation*. Springfield, NJ: Enslow Publishers, 1999.

> Discusses how companies use advertising and how it impacts people.

Dunn, John. *Advertising*. San Diego, CA: Lucent Books, 1997.

> Discusses the function of advertising and how companies use advertising to target the youth market. Discusses political advertising and the issue of freedom of speech.

Mierau, Christina B. *Accept No Substitutes: The History of American Advertising*. Minneapolis, MN: Lerner, 2000.

> Discusses the social history of advertising from 1600 to the present. Discusses how advertising has influenced and been influenced by American culture.

TOPIC 3. Dreamtime

In the mythology of the Australian Aborigines, Dreamtime was the legendary creation time when supernatural beings roamed the Earth and formed all of the physical features of the land. The concept of Dreamtime is complex because it defines both a mythological time period and a state of mind. Dreamtime existed in the distant past before the birth of human beings when only supernatural beings existed. In legend, because these supernatural beings created all that exists on Earth, the Aborigines consider these beings their ancestors. Therefore, Dreamtime exists in the minds of all who recognize the sacredness of the Earth and who feel an intimate connection with the land.

Modern day Aborigines who embrace the concept of Dreamtime believe that their ancestors remain on the Earth. The Aborigines also believe themselves to be an integral part of nature. They believe they have a sacred duty to guard and protect the land, and they do so in numerous ways. The ancestral spirits of Dreamtime formed the rocks and the rivers, the Moon and the Sun, and the mountains and the trees, and then they disappeared into their spirit homes and became plants and animals and forces of nature, such as the rainbow. In Aborigine thought, people today can tap into the power of the spirits by reenacting the myths and by engaging in certain ceremonies.

PROJECT IDEA

Have students choose one of the spirits of Dreamtime and create a poster that summarizes the myth of that spirit in pictures. Then have students explain their posters to the class. The sources in the Suggested Reading section contain myths of Dreamtime and of the supernatural beings who created the world during that legendary time.

SUGGESTED READING

Aboriginal Art Online. "Dreaming and the Dreamtime." *http://www.aboriginalart online.com/culture/dreaming.html.*

> Has a concise and easily understood explanation of Dreamtime and how Dreamtime creatures relate to the land.

Australian Museum. "Indigenous Australia." *http://www.dreamtime.net.au/dreaming/ storylist.htm.*

> Contains selected stories of Dreamtime and information on other aspects of Australian culture and spirituality. Has numerous links and curriculum guidelines for teachers and facts for kids.

Rahoorkhuit Network. "Eingana." *http://www.rahoorkhuit.net/goddess/goddess_quest/eingana.html.*

Has Australian legends, activities, and information on Australian art and culture. Has lots of information on the rainbow snake.

State Library of Queensland. "Aboriginal Resources for Young People." *http://www.slq.qld.gov.au/pub/festivals/aboriginal/young.htm.*

Contains resources on Aboriginal Australia for kids and teachers. Has booklists and links to Dreamtime stories and Web sites containing other cultural info on Dreamtime.

TOPIC 4. The Rainbow Snake

The rainbow snake is one of the most colorful and creative spirits of Dreamtime. This snake has appeared in Australian art in many forms.

PROJECT IDEA

Have students create a model of the rainbow snake. You can make it a classroom project and create a large snake to hang from the ceiling of your classroom to celebrate Dreamtime, like the Chinese dragons that often decorate classrooms and festivals to celebrate Chinese New Year. The books in the Suggested Reading section contain instructions for creating models from papier-mâché. You can make a papier-mâché paste simply by mixing flour and water and stirring until the paste is the consistency of thick cream.

SUGGESTED READING

The following two books contain instructions for creating projects with papier-mâché.

Capp, Gerry. *Great Papier Mache: Masks, Animals, Hats, Furniture.* Petaluma, CA: Search Press, Ltd., 1997.

Seix, Victoria. *Creating with Papier Mache.* San Diego, CA: Blackbirch Press, 2000.

PROJECT IDEA

Have students build a rainbow bridge out of colored tissue paper and florist wire. The bridge can be as large or as small as they'd like it to be. A class project might involve making a rainbow bridge so big that it arches over the doorway of the classroom. If you choose to do a large bridge, you might want to divide the class into seven groups and assign each group one color of the rainbow.

TOPIC 6. Color and Light

Nature makes beautiful rainbows, but people can create rainbows if they have sunlight and water. Remind students that to make rainbows, the Sun must be behind them, and the water must be in front of them.

PROJECT IDEA

On a sunny day, take students outside to make rainbows. All you need is access to a garden hose. Have students take turns holding the hose and making rainbows. Tell them to stand with the Sun behind them and spray the water in front of them. They should see a rainbow of colors in the spray of the garden hose.

—— SUGGESTED READING FOR TEACHERS

Boyer, Carl B. *The Rainbow: From Myth to Mathematics*. Princeton, NJ: Princeton University Press, 1987.

Discusses the science and mythology of rainbows.

Bromberg, Jeff. "Sunsets and Rainbows—Experiments for Kids." *Optics and Photonics News*. *http://www.opticsforkids.org/resources/Scattering_3.pdf*.

One of the many articles on the Optics for Kids site, a site that contains lots of articles and information for teachers on rainbows and atmospheric optics.

The Circle of the Dragon. "The Rainbow Serpent." *http://www.blackdrago.com/rainbowserpent.htm*.

Contains short myths of the rainbow serpent, primarily from Australia.

Greenler, Robert. *Rainbows, Halos, and Glories*. New York: Cambridge University Press, 1980.

Contains detailed explanations of reflection, refraction, and atmospheric optics. Contains computer simulations of optical phenomena.

Lee, Raymond L., Jr., and Alistair B. Fraser. *The Rainbow Bridge: Rainbows in Art, Myth, and Science*. University Park: Pennsylvania State University Press, 2001.

Has detailed explanations of the rainbow in art, science, myths, and cultures of different peoples. Contains illustrations of rainbows and symbolism.

Lynch, David K., and William Livingston. *Color and Light in Nature*. New York: Cambridge University Press, 1995.

Discusses and explains numerous optical illusions in the atmosphere.

Meinel, Aden, and Marjorie Meinel. *Sunsets, Twilights, and Evening Skies*. New York: Cambridge University Press, 1983.

Covers the nature of color in the atmosphere.

Myth*ing Links. "Floods and Rainbows: Mythology and Science." *http://www.mythinglinks.org/ct~floods.html*.

Explanations of rainbows in myth and science. Explains how rainbows relate to water and flood in myth and science and contains links to numerous related sites.

8 .... Auroras

THE MYTHS OF AURORAS

The aurora borealis, or the northern lights, looks spectacular in the night sky. In ancient times, it looked so spectacular that people told myths to explain its existence and offered fanciful explanations for why the lights appeared and what made them shine so brightly. Some people believed the northern lights appeared as an omen of flood, famine, or fire. Others believed they appeared as an omen of war. In the polar regions and in countries in the far north, the aurora borealis lights the sky in colors of green, red, purple, or blue. Many early myths assigned meaning to the phenomenon based on the colors they saw.

Any phenomenon that lit the sky so brilliantly made quite an impression on early people, and any phenomenon that appeared rarely and irregularly made it an object of curiosity. It was frightening to see the sky turn colors and not know how to explain it. It was also awe-inspiring to see the brilliant array of lights ripple and arc and swirl through the darkness. Many myths of the northern lights tell of animals or spirits dancing in the heavens, which explained why the lights appeared to move. Others tell of bloody battles in the sky world, battles fought by celestial warriors with shining weapons. The northern lights are both beautiful and frightening. Because they appear to move, many people feared that the lights could sweep down to Earth and chase people. Some people feared that the lights could pluck people off the Earth and take them up to the spirit world and hold them captive. People who did not fear the lights but simply delighted in their beauty imagined the aurora as dancing gods and goddesses. Some people imagined shining sleighs and horses carrying guests to a lovely wedding in the sky.

Read the following legend and discuss how it reflects early beliefs about auroras. Then use the section on science to explain the facts about auroras. A list of topics for discussion and projects follows.

"THE FIRE FOXES," A MYTH FROM FINLAND

It had been snowing for quite some time. It had been snowing for so long, in fact, that many of the animals had abandoned their summer homes and built dens underneath the ground. Snow covered the ground like a blanket and sparkled under the ice-cold sky. Inside the dens, the snow melted. It hollowed out a haven for foxes to shield them from the winter chill. The arctic fox was clever and quick, and he flashed like lightning across the snow-covered fells. Anyone who was watching could see the fox for just a second or two and then the sparks. Just before the fox disappeared beneath the ground, sparks of color flew from his tail and splashed into the sky.

Not everyone got to see these fabulous sights, of course, for they occurred far in the north where the Earth met the sky. But the Sami people knew these sights. The Sami were reindeer herders in Finnish Lapland, and they lived quite close to where the Earth met the sky. The Sami understood the ways of nature—they had to understand, for they lived in a hostile climate where it was bitter cold most of the year. The Sami also knew animals. They knew that the silvery white fox was spectacular, but they also knew that only a powerful silvery white fox could splash color into the sky. When the sky turned colors, it was Repo who did it. Repo the fox was so big and so strong that when he flicked the snowflakes with his tail, the entire sky turned ablaze with fire.

It was often difficult for the Sami to see the arctic fox scurrying across the fells, for the fox's coat was so white it blended into the snow. Furthermore, the fox moved quickly, so quickly that its bushy white tail flapped furiously against the snowdrifts. Sometimes, the fox got tired. The animal could no longer hold his tail up and keep it out of the snow. The Sami people watched the fox from a distance. They watched the fox's tail brush the snowflakes and send sparks into the sky. As the Sami watched, the sky turned flaming red. Moonlight bounced off the snowflakes. Sometimes, the sky blazed brightly in blue and green. When the sky ignited, the Sami people looked toward the heavens and watched for Repo. Many people thought they could see Repo in the winter. They could see him moving. Each winter the big white fox ran wildly most every night. He ran so far in the north that he touched the mountains. The people who watched the sky saw the movement. They watched the sparks of color ripple through the darkness and color the world with brilliant light. The people knew it was Repo who caused it. "It is foxfire we see," the people would say. And everyone who understood the ways of nature believed it.

THE SCIENCE OF AURORAS

"The Fire Foxes" is a myth that explains the Finns' perceptions of the aurora borealis, but it is also an expression of their belief in a sky world much like their own. Myths from northern Europe focus on the fiery forces that characterize the heavens and the contrasting forces that characterize the land. The Sami people of Finnish Lapland watched the animals in their world closely. They believed that the same animals that inhabited the world they knew inhabited the world in the sky.

Identify the elements in the myth that reflect early beliefs about auroras. Then compare these mystical elements to the science that emerged later.

BELIEF: Auroras come from the sky.

For a long time, people fortunate enough to see the aurora thought the spectacular lights were a sky phenomenon, but in fact auroras occur in the atmosphere. The lights originate from the Sun, though. During large explosions on the Sun, large quantities of solar particles are thrown into space in all directions. When these solar particles near the Earth, they get captured by the Earth's magnetic field, or the magnetosphere. When the particles hit the magnetosphere, they get pushed toward the poles. On their way to the poles, these solar particles are stopped by the Earth's atmosphere. They collide with atmospheric gases and emit photons, or light particles. When many collisions occur, people on Earth see auroras. Auroras are colored lights that seem to move across the sky.

BELIEF: Auroras come from the north.

Auroras inspired many different objects and characters in mythology, including a rainbow bridge in Norse myths, the dragon of Chinese myths, and a river of fire in Finnish myths. Most myths of auroras come from the polar region, where auroras are seen most frequently. Studies show that auroras occur most frequently over Finnish Lapland, where for a few months each year the Sun never rises very high in the sky. The people of Finland are accustomed to darkness in daytime, but during these periods of darkness the sky is frequently aglow with lights. People who live near the North Pole have many explanations to explain the lights. The lights seemed supernatural, and they seemed to originate from the northern sky.

For a long time, many people thought that auroras appeared only in the north, but we know now that auroras circle both the North and South poles. In the southern lands, they appear in the sky over the ocean. It was not until 1773, when the English explorer James Cook traveled to the South Pole, that people understood that the brilliant lights that appeared in the north appeared in the south at the same time. Both northern and southern lights have the same patterns, but they're reversed. It's simply much easier to see the northern lights because they shine over large areas of inhabited land. In the southern regions, the lights shine most clearly over the ocean far off the coast of Antarctica. The southern lights are called the aurora australis, and myths of the aurora australis circulate through the lands surrounding the South Pole.

BELIEF: Auroras appear in the winter.

In Finland, the northern lights occur most frequently in February and March and in September and October. They're actually seen least frequently in summer and in late December because the Earth tilts, and at these times of year the Earth's magnetic sphere is in the wrong position relative to the Sun. In summer, it's difficult to see the aurora also because the summer nights are light, but it is possible to see auroras at any time of year. Auroras are in the sky all the time, even in daylight, just like the stars are in the sky even in daylight. The reason we don't see the aurora and the stars in daylight, however, is because the Sun's light outshines them. Though it's possible for auroras to light the skies from evening until dawn, they are usually visible in the night sky for just one or two hours, usually close to midnight.

Take a look at the map of the North Pole in Figure 8.1. This map shows the area where the auroras are often visible. To determine the best times to see the aurora it is necessary to know the activity of the Sun and to monitor the solar cycle. This is what scientists do to study auro-

FIGURE 8.1 · The North Pole

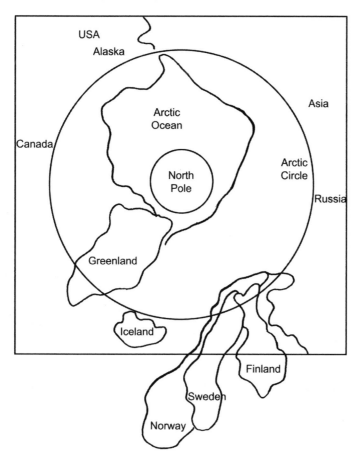

ras, and this is what photographers do to capture the beauty of an aurora display. Scientists and photographers have learned that to improve their chances of seeing a display they must first determine when solar storms will occur.

· ·

BELIEF: Auroras move.

People who have seen the aurora know that the lights bend and swirl and arc across the sky. To many people, the lights appear to be dancing. Some Finnish myths tell of the arctic fox dancing. Myths from other places tell of dancing animals of all sorts. The movement connected with auroras begins when solar particles leave the Sun. When great explosions, or sun storms, occur, debris from these sun storms shoots through space, riding on solar wind. *Solar wind* is a term astronomers use to describe clouds of plasma that travel through space. These clouds travel at around 217 miles a second. But even at this speed it takes them two to three days to reach Earth. When they do, they hit the magnetosphere, bend around the Earth, and collide with gases in the atmosphere near the poles.

BELIEF: Auroras are made of colored sparks.

When solar debris collides with gas molecules in Earth's atmosphere, the gases glow different colors. Sometimes the colors are so bright they extend from horizon to horizon. Sometimes it looks like the clouds are glowing. The brightness of the glow depends on the strength of the solar storm, and the color of the glow depends on which gases the solar wind hits when it collides with Earth's atmosphere. The atmosphere consists mainly of nitrogen and oxygen. When solar wind hits oxygen, auroras appear red; when solar wind hits nitrogen, they appear blue or green. The same thing happens in a neon light. In nature, altitude affects the color of the aurora. This is because different gases occur at different altitudes. Red light occurs at the highest altitudes, blue or violet light occurs at the lower altitudes, and green light occurs at the middle altitudes. When the Sun is particularly stormy, red lights occur even in lower latitudes. This is when people can see auroras far away from the poles. People far away from the poles are unaccustomed to seeing such spectacular lights, however, and they commonly mistake them for fire on the horizon. In theory, foxes could produce fire on the horizon because of the static electricity produced by the furry tail brushing against the snow.

BELIEF: Auroras have a connection to fire.

Galileo gave auroras their name in 1619. He named them after the Roman goddess Aurora, the goddess of dawn, because he believed the mysterious light to be a reflection of the dawn. Galileo was wrong about what caused the aurora, but he was right to connect them with the Sun. People all over the world connected fire with the Sun. The Finnish name for the lights (*revontulet*) comes from a Sami, or Lapp, legend, whereby the tail of a fox running along snow-covered fells strikes the snow drifts, sending a trail of sparks into the sky. *Revontulet* literally means "foxfire."

— TOPICS FOR DISCUSSION AND PROJECTS

TOPIC 1. The Beauty of the Aurora

Every February in Norway, the Andoya Rocket Range hosts a northern lights festival called Nordlyst. Festival-goers take lessons in both science and art. They create paintings of the northern lights, and they submit them to a jury. At the end of the festival, a jury chooses the best painting and awards the artist.

PROJECT IDEA

Have students create a picture of the northern lights using paints, crayons, charcoals, or any medium you can think of to capture the beauty of the aurora. Remind students that their picture might take the form of a realistic display, much like one of the photographs they see of the aurora, or their picture might represent a symbolic interpretation of the lights, inspired by one of the legends they read. Use the suggested readings to view photographs and pictures of the aurora and to get an idea of how impressive the lights look to those fortunate enough to witness a display.

SUGGESTED READING

Curtis, Jan. "The Aurora Page." *http://www.geo.mtu.edu/weather/aurora/images/aurora/jan.curtis/.*

> Has numerous photographs of auroras to give students ideas for their paintings.

Denlinger, Mish. "Auroras: Paintings in the Sky." *http://www.exploratorium.edu/learning_studio/auroras/.*

> Explains the science of auroras and has numerous links to additional information. Contains pictures of auroras to explain how they look from Earth and from space. Has a teacher page.

"The Night Rainbow" is a poem by Barbara Esbensen that incorporates images from aurora legends of many cultures. The poem features geese, whales, dancers, and raging sky battles. These were just some of the ways people explained the northern lights, but there were many other ways. People from Finland and other areas of Scandinavia connected the appearance of the lights to the weather. Commonly in Norwegian belief, the aurora was believed to be a harbinger of bad weather—particularly snow and wind. Some Inuit tribes believed the aurora indicated good weather.

PROJECT IDEA

View the following list of legends and beliefs of the northern lights. Write these legends and beliefs on the board or make a copy of this list to distribute in class. Have students create a poem that incorporates some of these beliefs. Then have students read their poems to the class.

Legends and Beliefs about the Aurora

- The Point Barrow Eskimos feared the aurora as evil, and they carried knives to protect themselves and keep the evil lights away.

- Native Americans from Wisconsin considered the aurora an omen of war. They thought the lights were the ghosts of their enemies who appeared in the sky to warn the people that they would seek revenge for their deaths.

- The Salteaus tribes from eastern Canada and the Kwakiutl and Tlingit tribes of Southeastern Alaska believed that the northern lights were the spirits of people dancing in the sky. Some of the Inuit tribes along the Yukon River believed that the aurora were the spirits of dancing deer, seals, salmon, and beluga whales.

- People of the Hebrides thought the lights were fairies, shining brightly and playing games in the sky.

- The Scottish and the Swedish people both believed the lights were dancers. The Swedes said the dancers were performing a polka or a folk dance.

- Some people in Canada called the lights puppets or marionettes.

- People from Estonia described the lights as a glow from a magnificent wedding in the sky. They imagined the horses and sleds of the wedding guests all shining radiantly.

- The Lakota Sioux thought the northern lights were children playing in the sky and waiting to be born on Earth.

- Some Inuit shamans believed they could use the aurora to cure disease. These shamans were thought to travel to the lights to get advice on how to heal the people.

SUGGESTED READINGS

Esbensen, Barbara. *The Night Rainbow*. New York: Scholastic, 2000.

> Celebrates the northern lights with images and legends of many ancient cultures. Includes notes about the legends and a scientific explanation of the lights.

Hall, Calvin, and Daryl Pederson. *Northern Lights: The Science, Myth and Wonder of the Aurora Borealis*. Seattle, WA: Sasquatch Books, 2003.

> Contains about 100 colorful photos of the northern lights, along with legends and science that explain the phenomena.

Souza, Dorothy M. *The Northern Lights: Nature in Action*. Minneapolis, MN: Carolrhoda Books, 1993.

> This book, in photo-essay format, contains an attractive collection of large color photographs and lots of information on the science and folklore of auroras.

TOPIC 3. The Mysterious Movement of the Northern Lights

Aside from the color and brilliance of the aurora, one of the most striking features of the phenomenon is how the lights appear to move. From their first appearance in the night sky until they disappear into the night, auroras seem to swirl, arc, and ripple. For this reason the northern lights awed and mystified ancient people and inspired them to tell stories to explain the movement.

PROJECT IDEA

Have students make a moving picture (a flip book) of the aurora. They can use a small pad of white paper and create their pictures using crayons, paints, or colored pencils or chalks. Tell them to paint a different appearance of the lights on each page. Students can use greens, blues, reds, and purples to show that auroras can appear in different colors during the course of one display. Students might choose to create a realistic picture of the lights, or they might prefer to create a symbolic picture. For those who would like to view pictures of the aurora, refer them to the following Web sites. Photographers have learned how to perfect their techniques to capture the movement of the aurora. Though many of the pictures on these Web sites were taken just seconds apart, each of the photographs looks entirely different.

SUGGESTED READING

Curtis, Jan. "The Aurora Page." *http://www.geo.mtu.edu/weather/aurora/images/aurora/jan.curtis/.*

> Has numerous photographs of auroras to give students ideas for their paintings.

Denlinger, Mish. "Auroras: Paintings in the Sky." *http://www.exploratorium.edu/learning_studio/auroras/.*

> Explains the science of auroras and has numerous links to additional information. Contains pictures of auroras to explain how they look from Earth and from space. Has a teacher page.

TOPIC 4. Exploring the Southern Hemisphere

Until explorers traveled to the South Pole, people wondered whether the Southern Hemisphere was simply an ocean of water or whether it contained another continent. James Cook was one of the first explorers to set sail for the South Pole. He discovered the southern lights, but Roald Amundsen was the first to reach the pole. Amundsen arrived at the South Pole on dogsled on December 14, 1911.

PROJECT IDEA

Have students write an entry in a travel journal from the perspective of James Cook, Roald Amundsen, or one of the other explorers who traveled to the South Pole. Students can read about these explorers in the following sources, and then use one significant aspect of their explorer's journey to write their entry.

SUGGESTED READING

Bramwell, Martyn. *Polar Exploration. Journeys to the Arctic and the Anarctic*. New York: DK Publishing, 1998.

> Discusses expeditions to the poles and the race for the South Pole. Also explains the land and climate of the poles.

Enchanted Learning. "James Cook: British Explorer." *http://www.enchantedlearning.com/ explorers/page/c/cook.shtml.*

> Discusses the explorations and discoveries of James Cook.

Enchanted Learning. "Roald Amundsen." *http://www.enchantedlearning.com/explorers/ page/a/amundsen.shtml.*

> Discusses the explorations and discoveries of James Cook.

Flaherty, Leo. *Roald Amundsen and the Quest for the South Pole*. New York: Chelsea House Publishers, 1992.

> Discusses Amundsen's journey to the South Pole.

Sipiera, Paul P. *Roald Amundsen and Robert Scott*. Chicago: Children's Press, 1990.

> Discusses the competition between Amundsen and Scott to reach the South Pole.

TOPIC 5. The Arctic Fox

In Finland in times past, the Artic fox could often be seen flashing across the fells and then disappearing quickly under the snow. Today, however, the Arctic fox is threatened in Finland and in other Arctic areas. If would be nearly impossible to catch a glimpse of the sight today, yet myths of the silvery white fox darting across the snow permeate Finnish myths and legends.

PROJECT IDEA

Have students write a report about the Arctic fox. They can write about the animal's behavior, habitat, physical characteristics, and status as an endangered animal.

SUGGESTED READING

Markert, Jenny. *Arctic Foxes*. Chicago: Child's World, 1991.

Matthews, Downs. *Arctic Foxes*. New York: Simon & Schuster, 1995.

SEFALO. "The Swedish-Finnish-Norwegian Arctic Fox Project." *http://www.zoologi. su.se/research/alopex/homesefalo.html.*

> Discusses the arctic fox and its endangered status in Sweden and Finland, and explains efforts to save the species.

TOPIC 6. The Arctic Climate

Winter is the longest season in Finland, and the temperatures are extreme. The temperature usually drops below freezing in mid-October, and in Lapland it can stay that cold for 200 days. Snow covers the ground all winter, the lakes freeze, and sometimes the Baltic Sea ices over completely.

PROJECT IDEA

Have students write a report about the climate of the Polar Regions. You may want to have them write a weather report for a week in the winter, or you may want them to give a weather broadcast to the class. You might want to assign each student a different country in the polar regions, or you might want to divide the class in two groups and have one write about the Arctic and one group about the Antarctic. The books in the Suggested Reading section are a few sources that may help students write reports about the weather in these cold climates.

SUGGESTED READINGS

Garrett, Don. *Scandinavia*. Austin, TX: Steck-Vaughn, 1991.

> Discusses the history, climate, geography, culture, and religion of Denmark, Norway, Sweden, Finland, and Iceland.

Miller, Debbie S. *Arctic Lights, Arctic Nights*. New York: Walker and Co., 2003.

> Explains the Arctic climate, the dark and light phenomena of the Arctic, and how animals adapt to the cold weather.

Riley, Peter D. *Survivor's Science in the Polar Regions*. Chicago: Raintree, 2005.

> Discusses the weather and climate of the polar regions and explains how to survive in emergency situations in cold climates.

Walsh Shepherd, Donna. *Tundra*. New York: Franklin Watts, 1996.

> Discusses the climate, land, and plant and animal life in the Arctic tundra.

Weller, Dave. *Arctic and Antarctic*. San Diego, CA: Thunder Bay Press, 1996.

> Discusses the land, climate, plant and animal life, and environmental concerns in the polar regions.

—— SUGGESTED READING FOR TEACHERS

Curtis, Jan. "The Aurora Page." *http://www.geo.mtu.edu/weather/aurora/images/ aurora/jan.curtis/.*

Has numerous photographs of auroras to give students ideas for their paintings.

Denlinger, Mish. "Auroras: Paintings in the Sky." *http://www.exploratorium.edu/ learning_studio/auroras/.*

Explains the science of auroras and has numerous links to additional information. Contains pictures of auroras to explain how they look from Earth and from space. Has a teacher page.

Eather, Robert H. *Majestic Lights: The Aurora in Science, History and the Arts.* Washington, DC: American Geophysical Union, 1980.

Provides an overview of the aurora in science, history, and myth.

Exploratorium. *http://www.exploratorium.edu/learning_studio/auroras.*

Contains curriculum-related materials on the northern lights.

Finnish Meteorological Society. "The Northern Lights in Finland." *http://www.fmi. fi/research_space/space_9.html.*

Explains the phenomenon of the northern lights in Finland. Has much about Finnish legends.

Hall, Calvin, and Daryl Pederson. *Northern Lights: The Science, Myth and Wonder of the Aurora Borealis.* Seattle, WA: Sasquatch Books, 2001.

Has about 100 color photos of the northern lights and much about the legends and science surrounding the phenomenon.

Jago, Lucy. *The Northern Lights: The Story of the Man Who Unlocked the Secrets of the Aurora Borealis.* New York: Knopf, 2001.

Discusses Norwegian physicist Kristian Birkeland, the man who studied the aurora borealis for years and was the first to connect the phenomenon to storms on the Sun.

NASA. *http://www.athena.ivv.nasa.gov/curric/space/aurora.*

Contains curriculum-related materials on the northern lights.

Savage, Candace. *Aurora: The Mysterious Northern Lights.* Richmond Hill, ON: Firefly Books, 2001.

Myth and science of the northern lights with lots of photos.

Shepherd, Donna Walsh. *Auroras: Light Shows in the Night Sky*. New York: Scholastic, 1996.

> Shepherd discusses what auroras are, where they are found, and how they occur. She also provides basic information on the various types, how scientists study them, and how to view and photograph them. Has legends and stories and lots of color photographs.

Virtual Finland. "Aurora Borealis II: Beliefs of Indigenous Peoples." *http://virtual. finland.fi/finfo/english/auroraborealis/aurora3.html*.

> Legends and beliefs about the northern lights.

Ytter, Harald Falk. *Aurora: The Northern Lights in Mythology, History and Science*. Herndon, VA: Anthroposophic Press, 1999.

> Discusses the beliefs and scientic theories of the northern lights throughout history and in different cultures. Has color illustrations.

Appendix: Teacher Resources

BOOKS

Andrews, Tamra. *Legends of the Earth, Sea, and Sky: An Encyclopedia of Nature Myths.* Santa Barbara, CA: ABC-CLIO, 1998.

Cavendish, Richard, ed. *Man, Myth and Magic: An Illustrated Encyclopedia of the Supernatural.* New York: Marshall Cavendish, 1994.

WEB SITES

This represents only a selective list of the numerous Web sites containing teacher resources about the weather. Most of the sites listed contain appropriate material for students in grades 4 to 8. Many also contain activities and experiments and/or links to other sites for information, activities, and lesson plans.

Dan's Wild Wild Weather Page. *http://www.wildwildweather.com/.*

> Contains weather information, lesson plans, and teacher resources for grades 1–12.

Earth and Sky. *http://www.earthsky.com.*

> *Earth and Sky* is a daily radio show that airs on hundreds of commercial and public radio stations nationwide. This is an excellent site for kids and teachers and contains a wealth of science facts, activities, and teacher resources. The site includes educational activities and classroom lessons, lecture notes, and virtual experiments, as well as many links to articles and other information.

How Stuff Works. *http://www.howstuffworks.com.*

How the Weather Works. *http://www.weatherworks.com.*

"How the Weather Works" is for teachers and students and offers a wealth of information on educational products to help you study the weather. Many of the products are for sale through the site, but there are also activities and lots of free information about rain and clouds on this site. Has excellent links to sites on many aspects of the weather and other sciences.

Internet4classrooms. "Earth Science." *http://www.internet4classrooms.com/earthspace.htm.*

This site contains a lengthy list of online sources for many areas of meteorology and atmospheric science as well as other areas of the earth sciences.

Kidspin. *http://www.producer.com/standing_editorial/kidspin/stories/science_library.html #thirtyseven.*

The science page of Kidspin has questions about science topics, many of them dealing with atmospheric science. Clicking on each question will give you an explanation of what happens and a simple experiment to do with students to show how it happens.

NASA. *http://spacelink.nasa.gov/Instructional.Materials/Curriculum.Support/Earth.Science/ Atmosphere.and.Weather/.*

National Center for Atmospheric Research. *http://www.ncar.ucar.edu.*

National Weather Service. *http://www.nws.noaa.gov.*

SCORE Science. *http://scorescience.humboldt.k12.ca.us/.*

A list of science resources on the Internet for K–12 teachers.

The Teachers Corner. *http://www.theteacherscorner.net/thematicunits/weather.htm.*

Designed for teachers of K–12 science, this site provides an extensive list of links to information and lesson plans on various aspects of the weather.

The Teacher's Guide: Weather. *http://www.theteachersguide.com/Weather.html.*

This site contains lesson plans and activities on numerous aspects of the weather. The activities are multidisciplinary and often introduce fresh approaches to old subject matter.

USA Today. "Guide to Science and the Atmosphere." *http://www.usatoday.com/weather/ resources/basics/wworks0.htm.*

Contains links to graphics and text on various aspects of weather phenomena.

The Weather Doctor. *http://www.islandnet.com/~see/weather/doctor.htm.*

This site is an incredible resource for learning the basics of atmospheric science. Easily accessible topics on numerous areas of weather phenomena, with lots of helpful links. Writing is rather advanced for young students but great for teachers searching for simple ways to explain the weather to students.

The Weather Unit. *http://faldo.atmos.uiuc.edu/WEATHER/weather.html.*

Has weather-related projects for teachers that tie climatology to math, science, social studies, language arts, geography, music, and PE. Includes ideas for reading and research, classroom tips, weather-related drama, and classroom props.

WeatherEye. *http://www.weathereye.org.*

Contains weather resources, lesson plans, activites, and projects for K–12 students.

Bibliography

Alth, Max, and Charlotte Alth. *Disastrous Hurricanes and Tornadoes*. New York: Franklin Watts, 1981.

Arnold, Caroline. *El Nino: Stormy Weather for People and Wildlife*. New York: Clarion, 1998.

Boyer, Carl B. *The Rainbow: From Myth to Mathematics*. Princeton, NJ: Princeton University Press, 1987.

The Circle of the Dragon. "The Rainbow Serpent." *http://www.blackdrago.com/rainbowserpent.htm*.

Crossley-Holland, Kevin. *The World of King Arthur and His Court: People, Places, Legends and Lore*. New York: Dutton, 1998.

Dolan, Edward F. *Drought: The Past, Present, and Future Enemy*. New York: Franklin Watts, 1990.

Donnan, John A., and Marcia Donnan. *Rain Dance to Research*. New York: David McKay Co., 1977.

Dunn, Andrew. *Fog, Mist, and Smog*. Austin, TX: Raintree Steck-Vaughn, 1998.

Flatow, Ira. *Rainbows, Curve Balls, and Other Wonders of the Natural World Explained*. New York: William Morrow, 1988.

Graham, F. Lanier. *The Rainbow Book*. Revised Edition. New York: Vintage Books, 1979.

Greenler, Robert. *Rainbows, Halos, and Glories*. New York: Cambridge University Press, 1980.

Heidorn, Keith. "Diamond Dust: Snow Without Clouds." *http://www.suite101.com/welcome.cfm/science_sky.*

Heuer, Kenneth. *Rainbows, Halos, and Other Wonders: Light and Color in the Atmosphere.* New York: Dodd, Mead & Company, 1978.

How Stuff Works. *http://www.howstuffworks.com.*

Jones, Ann. *Looking For Lovedu: Days and Nights in Africa.* New York: Knopf, 2001.

Kidspin. *http://www.producer.com/standing_editorial/kidspin/stories/science_library.html#thirtyseven.*

Krige, E. Jensen, and J.D. Krige. *The Realm of a Rain-Queen.* New York: Oxford University Press, 1943.

Lee, Raymond L., Jr., and Alistair B. Fraser. *The Rainbow Bridge: Rainbows in Art, Myth, and Science.* University Park: Pennsylvania State University Press, 2001.

Lindsay, David. "The Rain King." *Biography* 2(3), March 1998: 112.

Lynch, David K., and William Livingston. *Color and Light in Nature.* New York: Cambridge University Press, 1995.

Meinel, Aden, and Marjorie Meinel. *Sunsets, Twilights, and Evening Skies.* New York: Cambridge University Press, 1983.

Minnaert, M. *The Nature of Light and Colour in the Open Air.* Revised Edition. New York: Dover, 1954.

NASA. "Every Little Breeze Seems to Whisper Wind-Chill Factor." *http://observe.arc.nasa.gov/nasa/earth/wind_chill/chill_home.html.*

Neal, Philip. *The Greenhouse Effect.* London: B.T. Batsford Ltd., 1992.

Pettit, Don. "Auroras: Dancing in the Night" 2004. *http://earthobservatory.nasa.gov/Study/ISSAurora/.*

Pringle, Laurence. *Global Warming: The Threat of Earth's Changing Climate.* New York: Seestar Books, 2001.

Ramsey, Dan. *Weather Forecasting: A Young Meteorologist's Guide.* Summit, PA: Tab Books, 1990.

Reed, A.W. *Myths and Legends of Australia.* New York: Taplinger, 1965.

Roberts, Ainslie. *The Dreamtime: Australian Aboriginal Myths.* Sydney: Rigby, 1970. *http://www.mythinglinks.org/ct~air.html.*

Roberts, Jeremy. *King Arthur: How History is Invented.* Minneapolis, MN: Lerner Publications Co., 2001.

Sarapik, Virve. "Rainbows, Colours, and Science Mythology." *http://haldjas.folklore.ee/ folklore/vol6/rainbow.htm.*

The Weather Doctor. "Weather Phenomena and Elements." *http://www.islandnet. com/~see/weather/doctor.htm.*

Index

About the Author

TAMRA ANDREWS is an education writer and a language arts editor for Publishers' Resource Group and a former college reference librarian. She has written two encyclopedias of mythology, *Legends of the Earth, Sea, and Sky: An Encyclopedia of Nature Myths* (1998), and *Nectar and Ambrosia, an Encyclopedia of Food in World Mythology* (2000). Ms. Andrews developed a love of science and nature myths while working as an astronomy librarian for the McDonald Observatory at the University of Texas at Austin. She has written numerous scripts and articles for *Star Date*, an astronomy radio show and magazine produced by McDonald Observatory, and she has written various kinds of reference and educational materials for grades K–12. *Wonders of the Air* is the second in a series of four guides that teach lessons on natural phenomena through the reading of myths and legends. *Wonders of the Sky*, the first book in this series, also written by Andrews, was published by Libraries Unlimited in 2003.